Students' Edition

PROBLEMS IN ECOLOGY

M. K. Sands B Sc, M I Biol
Lecturer in Biological Education, University of Nottingham
Formerly Head of the Biology Department,
King Edward VI Camp Hill School for Girls, Birmingham

Mills & Boon Limited
London

OTHER PUBLICATIONS BY M.K. SANDS

Author of

Problems in Plant Physiology John Murray (1971)
Problems in Animal Physiology John Murray (1975)

Member of the Nuffield Advanced Biology Development Team, author and editor of some of the publications of *Nuffield Advanced Biological Science* Penguin Education (1970, 1971)

General Editor of Biology Topic Books

Human Populations (winner of The Times Educational Supplement Information Book of the Year Award, 1973)
Parasites
Space Biology Penguin Education (1972)
Published in hardback by Kestrel (1976)

Editor of

Understanding Biology Mills and Boon (1976)
Micro-organisms Mills and Boon (1976)

First published in Great Britain 1978 by Mills & Boon Limited, 17-19 Foley Street, London W1A 1DR.

ISBN 0 263 06352 6

Typeset by Reproduction Drawings Ltd., Sutton, Surrey.
Made and Printed in Great Britain by Thomson Litho Ltd., East Kilbride, Scotland.

CONTENTS

ACKNOWLEDGEMENTS

The author and publishers are grateful to copyright owners for permission to reproduce material, as listed below.

Table 1.3 Deevey, E. S., 1947. Life Tables for Natural Populations of Animals. *Q. Rev. Biol.,* **22**, 283–314. (Table 1)

Table 2.1 and Fig 2.6 Benson, W. W., 1972. Natural Selection for Müllerian Mimicry in *Heliconius erato* in Costa Rica. Science, **176**, 936–9. (From Tables 1, 2, 3 and Fig 1)

Table 2.2 Brower, L. P., *et al.*, 1964. Mimicry:differential advantage of color patterns in the natural environment. Science, **144**, 183–5. (Table 1)

Table 2.3 Park, T., 1954. Experimental Studies of Interspecies Competition: II Temperature, humidity and competition in 2 species of *Tribolium. Physiol Zool.,* **27**, 177–238. (Table 12)

Table 2.4 Harris, G. A., 1967. Some competitive relationships between *Agropyron spicatum* and *Bromus tectorum*. Ecol. Monogr., **37**, 89–111. (From Table 7)

Tables 2.5 and 2.6 Hawkins, H. S. and Black, J. N., 1958. Competition between wheat and 3-cornered Jack. *J. Aust. Inst. Agric. Sci.,* **24**, 45–50. (Table 2 and part of Table 3)

Table 3.1 Ricklefs, R. E., 1973. *Ecology*. Chiron Press Inc. New York, and Nelson, London. (Table 40.3) Based on material first published by Westlake, D. F., 1963, *Biol. Rev.,* **38**, 385–429, and Whittaker, R. H., 1970, *Communities and Ecosystems,* Macmillan, London.

Table 3.3 Edwards, C. A. and Heath, G. W., 1963. The role of soil animals in breakdown of leaf material. *Soil Organisms*, 76–84. Ed. Doeksen, J. and Van der Drift, J. North-Holland Pub. Co., Amsterdam. (From Fig 3)

Tables 4.1 and 4.7 Odum, E. P., 1959. *Fundamentals of Ecology* (2nd Ed). W. B. Saunders Co., Philadelphia and London. (From Tables 7 and 9)

Table 4.2 Hobson, T. (private communication).

Tables 4.3, 4.4 and 4.5 Bleasdale, J. K. A., 1960. Studies in plant competition. *Br. Ecol. Soc. Symp. 1, The Biology of Weeds*, 133–142. Ed. Harper, J. L. Blackwell Scientific Publications, Oxford. (Table II, data from Table 1 and p. 140)

Fig 1.4 Slobodkin, L. B., 1954. Population Dynamics in *Daphnia obtusa* (Kurz) *Ecol. Monogr.,* **24**, 69–88. (From Fig 2)

Fig 1.9 Clapham, W. B., 1973. *Natural Ecosystems.* Collier Macmillan Ltd., London, Macmillan Publishing Co., Inc. New York. (From Fig 3–23)

Figs 1.14 and 1.15 Bradley, B. C. (private communication).

Fig 1.17 Crisp, D. J., 1965. The Ecology of Marine Fouling. *Brit. Ecol. Soc. Symp. 5, Ecology and the Industrial Society*, 99–117. Ed. Goodman, G. T., Edwards, R. W. and Lambert, J. M. Blackwell Scientific Publications, Oxford. (Fig 2)

Figs 2.1 and 2.2 Huffaker, C. B., 1958. Experimental studies on predation: dispersion factors and predator–prey oscillations. *Hilgardia*, **27**, 343–383. (Figs 14 and 18)

Fig 2.3 MacLulich, D. A., 1937. *Fluctuations in the numbers of the varying hare* (Lepus americanus). *Univ. Toronto Studies, Biol. Ser., 43*. The University of Toronto Press, Toronto, Canada. (From Figs 16 and 17)

Figs 2.7 and 2.8 Gause, G. F., 1971. *The Struggle for Existence*. Dover Publications, Inc. New York. (Fig 25)

Fig 2.9 Crombie, A. C., 1946. Further experiments on insect competition. *Proc. Roy. Soc. (B)*, **133**, 76–109. (Figs 6 and 9)

Fig 2.12 Vaurie, C., 1951. Adaptive differences between two sympatric species of nuthatches (*Sitta*). *Proc. X Int. Ornithol. Congress*, 163–6. Ed. Hörstadius, S. (Fig 3)

Fig 2.13 Lack, D., 1947. *Darwin's Finches*. Cambridge University Press (From Fig 17)

Fig 2.14 Donald, C. M., 1961. Competition for light in crops and pasture. *Symp. Soc. Exp. Biol. XV, Mechanisms in Biological Competition*, 282–313. Ed. Milthorpe, F. L. Cambridge University Press. (Fig 3)

Fig 2.15 Palmblad, I. G., 1968. Competition in experimental populations of weeds with emphasis on the regulation of population size. *Ecology*, **49**, 26-34. (From Table 6)

Fig 2.16 Grümmer, G., 1961. The role of toxic substances in the interrelationship between higher plants. *Symp. Soc. Exp. Biol. XV, Mechanisms in Biological Competition*, 219-228. Ed. Milthorpe, F. L. Cambridge University Press. (Data from pp. 226–7)

Figs 4.3 and 4.4 Walpole, P. R. and Morgan, D. G., 1972. Physiology of grain filling in barley. *Nature, Lond.*, **240**, 416–7. (Fig 1 and part of Fig 2)

Fig 4.5 Williams, J.I. and Shaw, M., 1976. *Micro-organisms*. Mills and Boon, London. (From Fig 8.7)

Fig 5.9 Burgess, E., 1976. Life on Mars creeps closer. *New Scientist*, **71**, 480. (Fig 1)

Figs 6.2, 6.3 and 6.4 Hynes, H. B. N., 1959. The biological effects of water pollution. *Inst. Bio. Symp. 8, The Effects of Pollution on Living Material*, 11–24. Ed. Yapp, W. B. (Figs 2 and 3)

Figs 6.5 and 6.6 Gilbert, O. L., 1965. Lichens as indicators of air pollution. *Brit. Ecol. Soc. Symp. 5, Ecology and the Industrial Society*, 35–48. Ed. Goodman, G. T., Edwards, R. W. and Lambert, J. M. Blackwell Scientific Publications, Oxford. (Fig 2, and based on Fig 1)

Fig 6.9 *National Atlas of Disease Mortality in the U.K.* Royal Geographical Society. Ed. Howe, G. M. Nelson, London, 1970 (2nd Ed). (From map p. 59)

Fig 6.10 *Smoking and Health*. A report of the Royal College of Physicians. Pitman Medical Publishing Co. Ltd., London, 1962. (Fig 11)

Fig 6.11 Bradshaw, A. D., 1971. Plant evolution in extreme environments. *Ecological Genetics and Evolution*, 20–50. Ed. Creed, R. Blackwell Scientific Publications, Oxford. (Fig 2.12)

Fig 6.13 Moore, N. W., 1967. A synopsis of the pesticide problem. *Adv. Ecol. Res.*, **4**, 75–130. Ed. Cragg, J. B. Academic Press (London) Ltd. (Fig 5)

Problem 1.6 Data from Swanson, H. (private communication)

The author wishes to record thanks to the following for their assistance: G. S. Bailey, P. E. Bishop, Dr A. Booth, S. W. Hurry, K. E. Selkirk, J. B. Wightman. The author also wishes to thank the following teachers and their sixth-form students who tackled the problems in their initial form:

G. S. Bailey, Sherwood Hall Upper School, Mansfield.

S. M. Baldwin-Wiseman and B. Bottomley, West Bridgford College of Further Education, Nottingham.

P. E. Bishop, The Brunts School, Mansfield.

I. R. Whitehead, The Holgate School, Hucknall.

1

POPULATIONS

1.1 Population growth

A **population** consists of all the individuals of the same species found occupying a given space. A population can start with only a few individuals, or even one, and it can grow to utilize the available resources. At some stage in its growth there will be **checks** imposed by the environment, setting a limit to the size of the population. The checks may cause it to decline, crash, and maybe even disappear. Fortunately, the potential growth of a population is never realized. A single oyster can produce one million eggs a year, so the population has the capacity to grow at an enormous rate. If all the offspring reproduced once, after five generations there would be 10^{30} oysters, weighing something like 500 times the mass of the sun. Suppose a female housefly in a kitchen lays 120 eggs, and there are seven generations of houseflies per year. The final number could be something like 5×10^{12} (5 million million) flies after one year.

If a small population of a species is introduced into a new area which is suitable for its growth, one sees the beginning of a **population explosion**. An example is the introduction of about 200 000 sheep to the island of Tasmania in 1820. By 1850 there were about two million sheep. The population had more than doubled every ten years, and up to then there had been no checks on the growth of the population.

Part 1

1 What do you think population growth depends on? What sort of checks may be imposed on its growth?

The growth curve for the early stages of population growth for most species of animals and plants is sigmoidal, producing a growth curve which can also often be applied to the growth of an individual or an organ. A study of populations in the laboratory has given a great deal of data on population growth. Examples of organisms which could be studied in a school laboratory are a yeast (*Schizosaccharomyces pombe*), a bacterium, or a unicellular alga.

In one school investigation, several flasks half-filled with a culture medium and then sterilized were inoculated with approximately the same number of yeast cells. The cultures were left at 30 °C. Samples were removed from the flasks at intervals and the number of cells per unit area of a counting chamber was counted using a microscope.

2 Why were the flasks sterilized?

3 Why was a number of flasks used instead of just one flask?

Table 1.1 Growth of a population of yeast

Hours after inoculation	0	2	4	6	8	10	12	14	16	18
No. cells counted per unit area	10	30	70	175	350	515	595	640	655	660

The results from this experiment, as a mean for all flasks, were as shown in Table 1.1.

4 Construct a graph of these results, putting time on the horizontal axis.

5 The graph shows you the typical sigmoidal curve of population growth. The distinct parts of the curve are called the **lag phase**, the **logarithmic** or **exponential phase**, and the **stationary phase**. Which part of the curve do you think is each of these? For the laboratory culture just described, say what you think is happening in each phase.

6 What differences in the results would you have expected if

a the flasks had been incubated at a temperature
i 10 °C lower than that used, **ii** 10 °C higher than that used

b the concentration of the nutrient medium had been
i double that used, **ii** half that used?

7 If a curve such as this is used as a model for population growth, what assumptions are made?

8 Now construct a curve which shows the **rate of population growth** in this laboratory investigation. Do this by subtracting each number from the next and plotting the figures so obtained against time. Such a curve shows more clearly than the sigmoidal population growth curve how quickly or slowly the population is increasing. Where is the turning point of population growth in this example?

Part 2

Population growth is often plotted on a logarithmic scale because of the vast numbers involved. Figs 1.1 and 1.2 are for comparison. Fig 1.1 shows a hypothetical curve obtained plotting numbers of bacteria per cm^3 of culture medium. The population doubles every hour for the first eight hours. At six hours the curve runs off the graph. Fig 1.2 shows the population growth expressed semi-logarithmically.

Fig 1.1 Population growth of bacteria

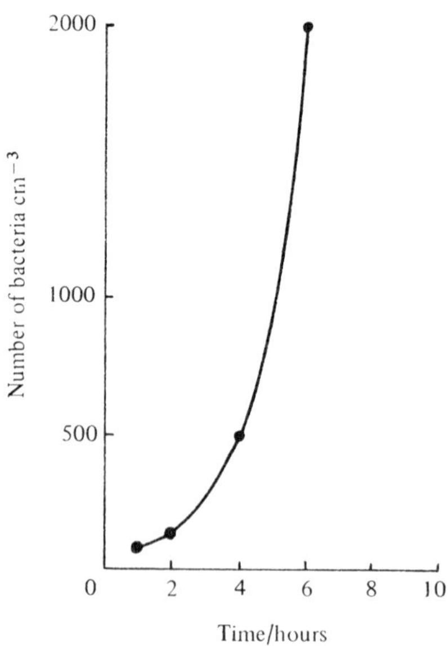

Fig 1.2 Growth curve for bacteria

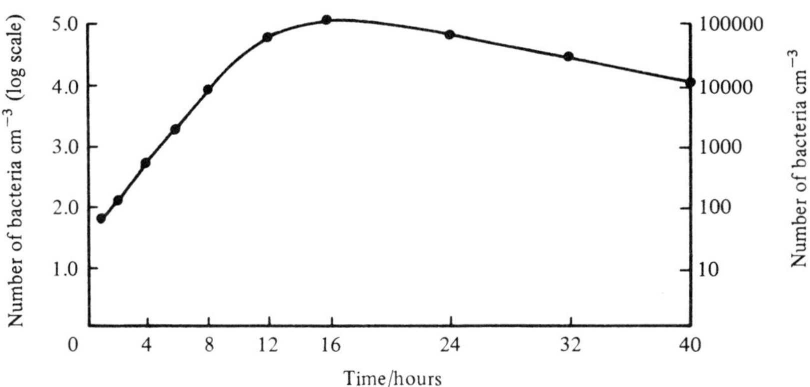

9 Which point on Fig 1.2 corresponds to the place in Fig 1.1 where the curve runs off the graph?

10 How many bacteria were present per cm^3 when the rate of population increase began to slow down?

Part 3

Some populations, instead of levelling off, have a steep decline when many individuals die at about the same time. The growth curve has a **J** shape. An example is shown in Fig 1.3 for a unicellular alga of British lakes. The alga undergoes very rapid seasonal growth to produce what is called a **bloom**. There is rapid, unrestricted, exponential growth up to the limits of the environment. As nutrients are exhausted, or the weather changes, or toxic wastes accumulate, the population dies off.

Fig 1.3 A J-shaped growth curve for an alga

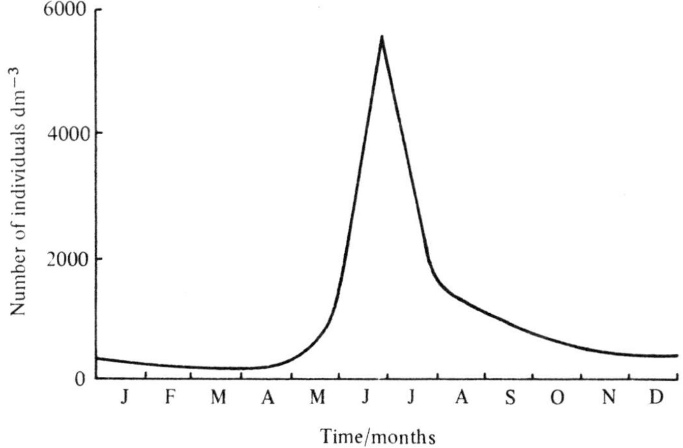

11 In what sort of organisms do you think a J-shaped curve could be observed?

Part 4

Animals with a more complex life history than those already mentioned may at first give a sigmoidal curve as the population grows in a new area, but thereafter there can be oscillations, sometimes quite violent ones. As just one example, consider Fig 1.4 which shows the total numbers (adults and young) of asexually reproducing *Daphnia* in a laboratory population over about one year. The population started with a single female *Daphnia* carrying eggs in a transparent sac on her back. The environmental conditions were constant, the amount of food supplied (unicellular algae) was constant, and there were no other living organisms in the culture apart from the *Daphnia* and the algae.

Fig 1.4 *Daphnia* population curve

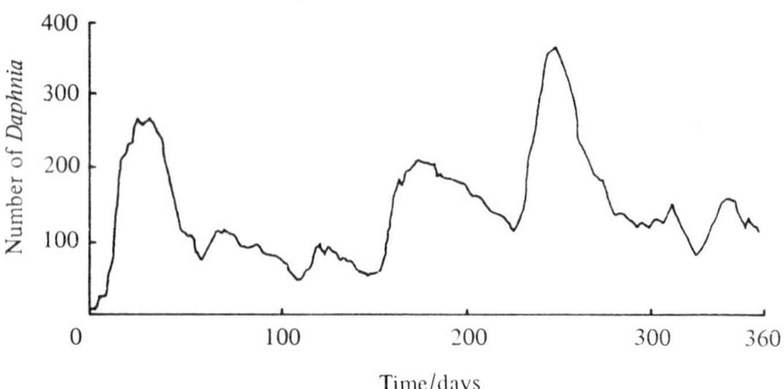

12 Attempt to explain the oscillations in this curve.

1.2 Population growth under different conditions

Figs 1.5, 1.6 and 1.7 show the growth of populations of three organisms, each cultured under a set of conditions where one variable was changed. For the yeast,

Fig 1.5 Growth of populations of yeast in different conditions

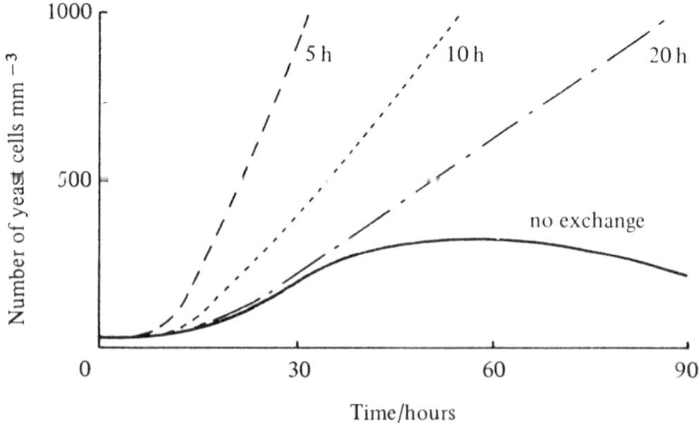

Fig 1.6 Growth of populations of *Daphnia* at two temperatures

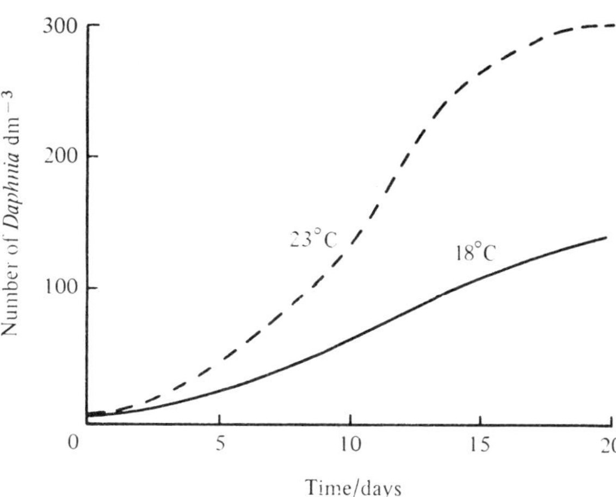

Fig 1.7 Growth of populations of flour beetles with different amounts of food

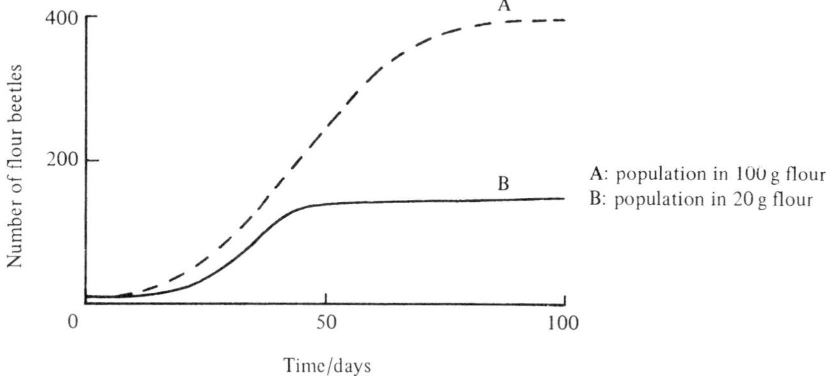

the medium in and on which it was growing was exchanged for a fresh medium every so many hours, as indicated against the curves on Fig 1.5. For the *Daphnia*, the temperatures were different for the two populations. For the flour beetle the amount of food available, in which the beetles lived, was different for each of the two populations.

Discuss each of these three sets of curves, saying how you think the changing environment affects the population growth.

1.3 Human population

We have become familiar in recent years with articles on the increasing number of people in the world and the potential threat of overpopulation. The increase in numbers is estimated at between 70 and 80 million per year, an annual growth rate of 1.9%. If you feel your pulse you can reckon on almost two babies born for each pulse beat.

Table 1.2 gives figures for the number of people in the world based on a number of sources, mainly various United Nations publications. The figures for recent years are calculated from census counts and written records of a more or less reliable nature. Figures before the year 1800 are estimates, based on whatever evidence is available or on historical guesses. Those for pre-history are informed guesses and vary somewhat from authority to authority.

Table 1.2 Approximate estimated world population of people

BC	1 000 000	125 000
	300 000	1 000 000
	25 000	3 340 000
	8 000	5 320 000
	4 000	86 500 000
	3 000	150 000 000
	0	250 000 000
AD	1000	350 000 000
	1650	500 000 000
	1750	730 000 000
	1800	800 000 000
	1850	1100 000 000
	1900	1600 000 000
	1950	2500 000 000
	1960	2900 000 000
	1970	3600 000 000
	1971	3670 000 000
	1975	3967 000 000
	2000	c 7000 000 000

1 With time on the horizontal axis plot these figures, or as many as you can accommodate, on to two graphs:

a a normal graph
b a logarithmic graph, both axes using a log scale.

2 The second graph shows rapid increases in the size of the human population at three times in history and pre-history. Say what kinds of development in the human way of life you think these three increases reflect.

3 Consider graph **a**, showing a massive increase in numbers over the last century and a greater probable continuing increase into the future. If this graph showed a population of animals in the wild, what natural forces or checks do you think would soon bring the population down?

In human populations such checks have, of course, happened at various times. In many parts of the world there are hunger, malnutrition and famine even today, often accompanied by diseases such as cholera and typhus. One and a half million people died in the single Indian state of Bengal in 1943 due to famine. The Irish famines of 1845 and 1846 caused about one million deaths and more than a million people to emigrate.

The main cause of death, however, has always been disease. Malaria, until recently, was a great scourge of mankind. Before 1946 there were 300 million cases of malaria a year, with 3 million deaths. With the massive use of insecticides, malaria has been reduced to 120 million cases a year, of which 100 million are in tropical Africa. In the fourteenth century plague carried away a quarter of Europe's population. Before the plague, the population of Britain had reached four million from the one million recorded in Domesday Book. It was down to about two million in a few years and took nearly 300 years to reach four million again. Infectious diseases used to take a heavy toll, particularly amongst children. Today, however, the killer diseases in industrial countries are not the infectious ones but cancer, heart disease and various diseases of old age.

Interspecific competition (competition between different species) is hardly a check on human population growth, but none-the-less a frightening amount of food is spoiled in storage by pests which compete with us, quite favourably to themselves. In addition, our livestock and crops are consumed or attacked by pests, pathogens, consumers, predators of one sort or another.

Intraspecific competition (competition between members of the same species) seen in warfare, even in infanticide and abortion, motor accidents, or in crime, can account for large numbers of deaths and thus be a check on population growth. About twenty-two million people, soldiers and civilians, died in five years during the second world war, nine million of them Jews in the gas chambers and other Nazi pogroms, a most dreadful amount of human suffering. But even this vast number is such a small part of the total world population, and more babies were born than normal (the bulge) after the war, that the loss is hardly marked on a graph such as that drawn for question **1**. Indeed, in a world population which is increasing at the rate of 70 million per year, twenty-two million of these extra births would occur in about four months.

Conscious individual choice on the number of children and when to have them is a recent feature. We shall increasingly rely on it as improved medical and environmental technology continues to reduce the mortality rate. In some countries, laws and the choice of society control the population to a certain extent. In Japan, for example, there are financial incentives through the state for fewer children and also legalized abortion.

Space and climate may have set a limit to man's increase in the early days of our evolution, when a large area was needed to support a food-gathering and hunting family. Space or other resources may well prove a limiting factor in the future.

4 The recent growth of the human population gives rise to a lot of speculation about the shape of the curve to come. Natural populations show first a sigmoid growth curve and then one or more of a number of possible patterns: a steady equilibrium, oscillations, a J shape (Problem 1.1). Is the human population graph a sigmoid curve still in its log phase? Is it a J-shaped curve which will collapse soon or eventually? Will it move into a phase of steady equilibrium or will it show oscillations and fluctuations?

Consider any factors which could cause each of these patterns to happen in the future.

5 Earlier increases in the human population of the world have probably gone hand in hand with an increase in the **carrying capacity** of the environment. Carrying capacity is the size of the population the environment can support. The increased carrying capacity may have resulted from new territory being made available (by means of travel, perhaps by changes in climate) or greater efficiency in food production. We do not yet know at what level the human curve will be forced to stop rising. The carrying capacity of the world could be increased by our own efforts, or we could exceed the carrying capacity for some time, like the deer on the Kaibab Plateau (Problem 2.1). In what ways do you think the carrying capacity could be increased?

6 We seem about to double the world's population every thirty or so years. What are the consequences or implications of this doubling for the quality of life for everyone?

1.4 Populations and age

The size of a population and whether it increases or decreases depends mainly on the **natality** (birth rate, or rate at which new individuals enter the population), **mortality** (death rate), the length of life of individuals, emigration and immigration. The distribution of ages in a population is related to the growth rate of that population. The age distribution can be used to estimate whether the population is expanding, contracting or stable. One way to show the age distribution is by an **age pyramid**: a vertical bar graph which shows the number or proportion of individuals in different age groups at any one time. The youngest are at the bottom and the oldest at the top, males on the left, females on the right. The age pyramid shows the age distribution at any one given time. As the population age distribution changes with time so, of course, does the form of the age pyramid. Both natality and mortality are affected by the age distribution. Reproduction is usually restricted to certain age groups and mortality usually varies with age, probably being highest in the very young and the old.

1 How, then, would an age pyramid allow a prediction to be made about the future population?

Fig 1.8 shows stylized diagrams (not bar graphs) of three types of age pyramid.

Fig 1.8 Age pyramids

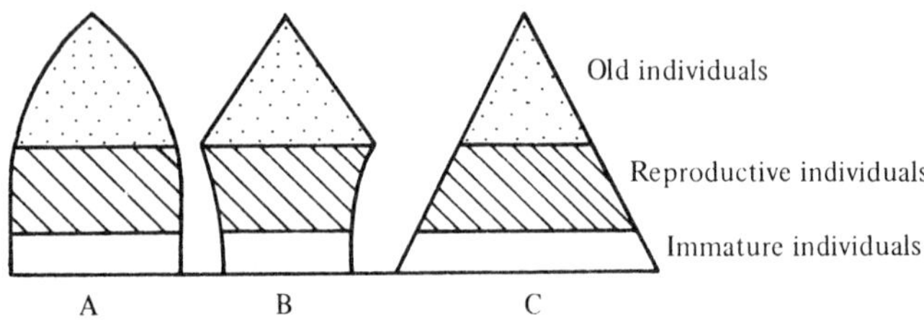

2 Match the diagrams in Fig 1.8 to the following statements:

a a rapidly expanding population
b a stable population
c a declining population.

Fig 1.9 Age pyramids for human populations in three countries

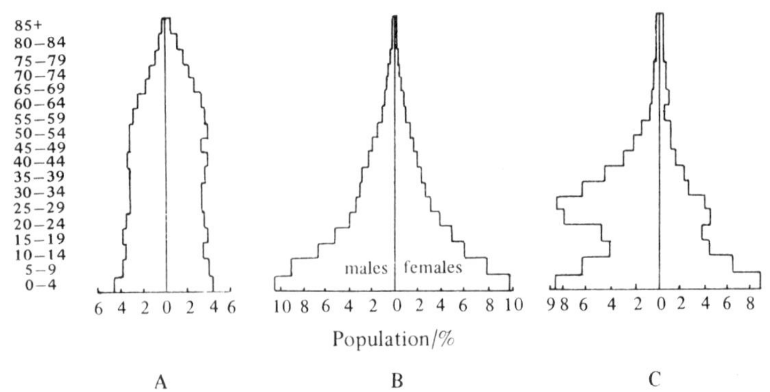

You can apply the same principles to age pyramids of human populations. An age pyramid of one country may be different from that of another. You can also see the effects of national or international events on the population. Fig 1.9 gives three age pyramids. One is of a South American country (Costa Rica, 1963) with a rapidly growing population; one of the UK (1965) an industrial country with a stable population; and one of Kuwait (1965), a state on the Arabian Gulf which has grown rapidly in the past few decades, has welcomed substantial immigration from other Arab countries, and is now one of the richest countries in the world due to its oil.

3 Say which pyramid represents the population of

a Costa Rica, b UK, c Kuwait.

Fig 1.10 Age pyramid for East Germany 1960

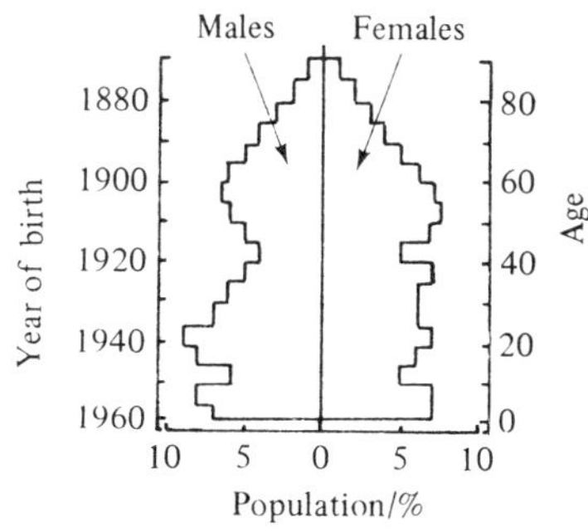

4 Fig 1.10 shows the age pyramid for East Germany in 1960. On this diagram point out the effects of

a a reduced birth rate during world war I (1914–18),
b a reduced birth rate after world war II (1939–45),
c war deaths, especially for males, during world war II.

A country such as Costa Rica (or many countries in South Asia, Africa and South America) with a rapidly expanding population, may eventually change from a pyramid with the weight at the young end to a bell-shaped pyramid indicating a more stable population.

5 What social and medical trends within a population would help to change the shape of the pyramid in this way?

6 Discuss the social effects on a population of having

a a broad-based pyramid, b a top-heavy pyramid.

1.5 Fertility

Some factors which affect the **fecundity** (the reproductive capacity) of a species are the number of offspring produced at any one birth-time, the length of time between births, the length of time for which the female is physiologically able to reproduce, the proportion of adult females which produce offspring, and the **generation time** (the time from the birth of an individual to when that individual reproduces). Some factors which determine how many young reach reproductive age are disease, predation, and shortage of food leading to death by starvation.

1 Consider the town or place where you live and say which, if any, of these

factors have caused an increase or decrease in the population in the past one hundred years or for any other suitable period of time.

Fertility (the number of offspring actually produced by an individual) is not usually constant for each age group within a population. Many species have a reproductive period, before and after which they cannot reproduce. In others, fertility is related to the size of the organism and it increases throughout life. The age distribution of fertility can have an effect on a population's birth rate and thus its size.

2 Consider two human populations of 1000 people, one in a country such as the UK, and one in a country with a high birth rate. In the UK the average number of children per family has recently rapidly gone down to 2.1. In the other country it could be as high as 4 or more. In the UK about one-quarter of the population is between the ages of 20 and 40 when most reproduction takes place. In the other country let us say that 30% of the population is between the ages when children are usually produced, which is probably a younger age range than 20–40 years. Calculate the possible size of each population after one generation.

3 Is generation time important? Consider the contributions made to the future population by two women. One has twins when she is 19 years old and has her third and last child at 21. The other starts her career and becomes established in it, marries late and has five children, all born after she has passed the age of 33. Which woman contributes more to the reproductive potential of the population?

4 What effect do you think equality for women will have on the reproductive potential of a population?

1.6 Population density and stress

Golden hamsters can breed quickly and soon produce a large population. The young are sexually mature at about six weeks, the oestrous cycle is four days, the gestation period sixteen days, and the litter size between four and twelve young. There is thus a large reproductive capacity. A single pair of golden hamsters could theoretically quickly cause a population explosion. Indeed, all laboratory and pet golden hamsters in the UK are descendants of a litter imported from Syria in 1930.

The population growth of the golden hamster in the laboratory was investigated. Pairs of hamsters were put into each of eight enclosures (floor space of 80 cm × 90 cm), with a nest box and nesting material. There was always a good supply of food and water. In such an investigation external factors which usually act to limit the population such as predation, disease, lack of food, bad weather and the possibility of emigrating, are all absent. The investigation continued for eight months, and was repeated six times. As there were eight enclosures each time there were 48 replications.

In all enclosures, over the whole length of the investigation, a total of 56 pregnancies was recorded. Only about half produced live litters, in the remaining pregnancies the young were either aborted or resorbed. Of the live litters, only six litters survived to maturity, the others all died within three days of birth. The dead young, possibly killed by the mother or other adults, or dead from other causes, often disappeared. They were presumably eaten, although the investigators did not actually see any cannibalism. The largest number of hamsters in any one enclosure at any time was nine. The population had stabilized at a low level.

1 Suggest one or more reasons for the behaviour which resulted in such a low population. Consider sexual, maternal and social behaviour.

2 How would you re-design the investigation to avoid the design which possibly caused so much disruption of normal behaviour?

1.7 Surviving

Within a population, an individual can look forward to a life of a certain length, based on the trends in that population. This is called the **life expectancy**, and is the average duration of life of all members of that population. As far as human populations are concerned, some people obviously live to far beyond the average life expectancy and some die well before it.

Ancient Egyptians had a life expectancy of only 22 years. In Roman times the life expectancy was about 30 years. A hundred years ago a boy born in England could expect to live to age 41, and a girl to 45. By 1900 the life expectancy had crept up to between 45 and 50 years. Now it is between 70 and 75 years. Baby girls in Norway have the world's longest life expectancy: 77.6 years. For most African countries the figure is less than 50 years. For other animals life is usually cut short by accident or disease, starvation or predation. Amongst mammals, elephants come next to man. The oldest recorded elephant was in a zoo, where it died at the age of 67 years, while bears can live for up to about 30 years and lions for 20.

The lower life expectancy figures for humans were due to the very heavy mortality amongst children. In this country a hundred years ago, four babies out of every ten died before they reached adulthood. In 1900 the figure was 162 per thousand. Nowadays twenty-nine infants out of every thousand born die before their first birthday. In many countries in the world today the infant mortality rate is even higher than our 1900 rate. But even in Roman times the life expectancy of an 80-year-old person was about the same as that of an 80-year-old today. In both, the body is running down and death is likely to occur from 'old age', that is, from a degenerative disease or the inability to repair body damage. Females live longer than males in other species as well as in humans. In humans, males seem to be inherently more susceptible to death as well as having greater exposure to hazards through occupation or environment.

A **survivorship curve** can be plotted for a population on the basis of survivors per thousand (say) against age, or percentage of maximum life span achieved. To construct a survivorship curve a total population of individuals, such as a thousand, is

Fig 1.11 Survivorship curves

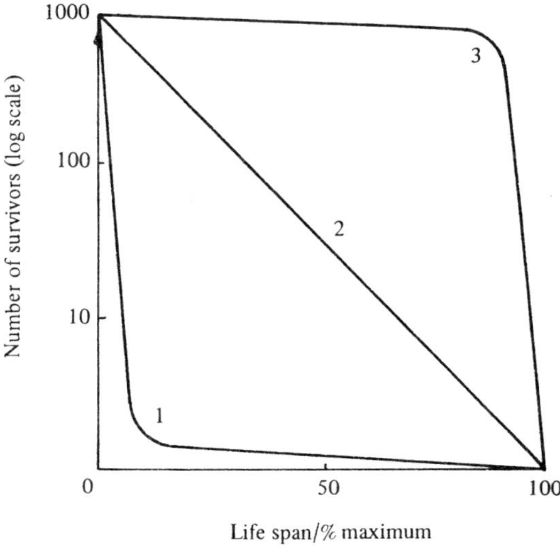

considered at birth. At regular intervals of time, the total number of survivors from this thousand is plotted, and the curve drawn. Some populations, such as an oyster population, or most fish, will show a quite different curve from that of present-day man in a developed country with a stable population. Oysters produce a vast number of young, most of which die. The individual oyster becomes relatively safe only when it settles. Western man, on the other hand, has a long life expectancy. Fig 1.11 shows stylized survivorship curves for three types of population.

1 Match each of the following statements with one of the curves in Fig 1.11, and give examples of animals which you would expect to fit each curve.

a Mortality is concentrated in older animals and the possibility of death increases with age. This population probably has very few, if any, natural enemies.

b Most mortality is in the very young stages; the probability of death decreases with age.

c Mortality decreases gradually with the passage of time and the probability of death is constant with age.

2 What factors over the last century or so have worked together to produce a curve for man such as curve 3 in Fig 1.11?

3 Draw two graphs, and compare them, to illustrate what you consider to be the life expectancy of people living in

a the UK,

b an underdeveloped country with few resources or social services.

The shape of the survivorship curve can vary with the number of individuals in the population and how crowded they are. As an example consider Fig 1.12. Each curve represents a stable deer population living in Scotland. The most crowded population (about twenty deer per square kilometre) lived in an area managed by man. The open heathland of heather and shrubs was maintained by burning from time to time, giving new growth which provided more food. The survivorship curve for this population is curve A on Fig 1.12. The more widely spaced population of about seven per square kilometre was in an unmanaged area of old bushes and tall shrubs, (curve B).

Fig 1.12 Survivorship curves for two populations

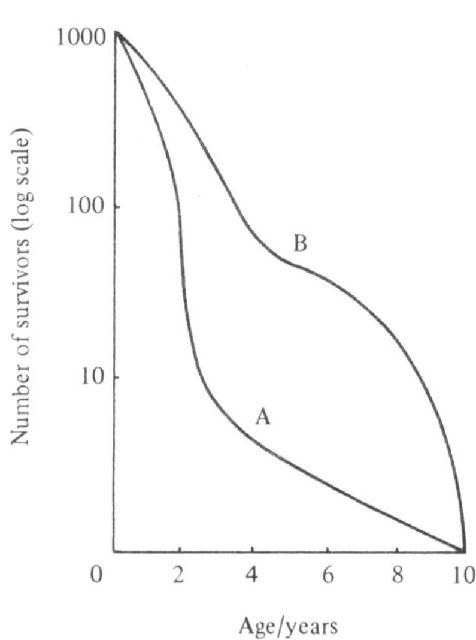

4 Which population of deer has the longer life expectancy?

5 Why? (Think why some men should want to manage a deer population on a Scottish moor.)

1.8 Survival in mountain sheep

The data in Table 1.3 were collected in 1937. The table relates to the length of life of Dall mountain sheep, found in a national park in Alaska. The age of the sheep can be determined from the number of rings on the horns which remain intact for a long period after death. The natural predators of the sheep were wolves, although many would die from other natural reasons. They were not killed by man.

A **life table** was drawn up from 608 skulls or horns collected over several years. They were considered to be a representative sample of the population.

(Life tables for people are used by insurance companies when calculating the risk of a life they insure. The risk affects how much the person pays for the insurance. A life table shows, for a group or cohort of individuals, the mortality rate for each age group, the numbers surviving and their life expectancy. Which sex and what ages would pay most for their insurance?)

Table 1.3 Life table of the Dall mountain sheep

	Out of 1000 born	
Age in years	Number dying in the age interval	Number surviving at the beginning of age interval
0–0.5	54	1000
0.5–1	145	946
1–2	12	801
2–3	13	789
3–4	12	776
4–5	30	764
5–6	46	734
6–7	48	688
7–8	69	640
8–9	132	571
9–10	187	439
10–11	156	252
11–12	90	96
12–13	3	6
13–14	3	3

1 a Plot a survivorship curve for the sheep using either number surviving per thousand against age (on horizontal axis), or log number survivors against life span as a percentage of total lifespan (as in Fig 1.11).

b Fig 1.11 (Problem 1.7), shows three survivorship curves. Which of these three is similar to the one you have drawn?

2 Which are the most precarious years of life for these sheep, and why?

3 Fig 1.13 gives the life expectancy curve for the Dall mountain sheep. Use the graph to give the life expectancy of the sheep at

1 year, 5 years, 9 years, 12 years.

Fig 1.13 Life expectancy curve for mountain sheep

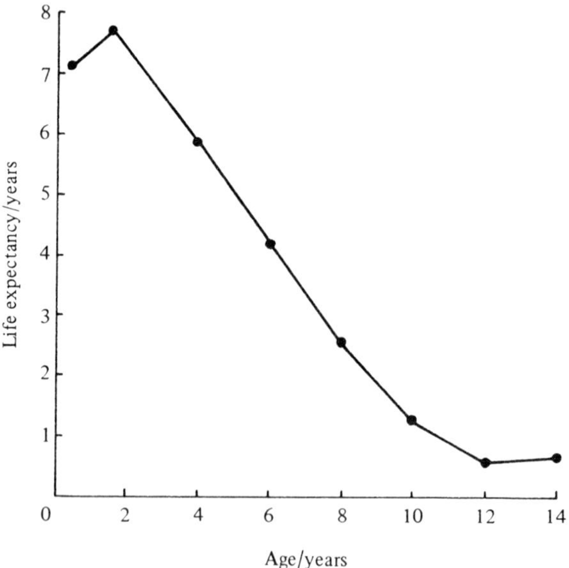

4 What is the mean life expectancy?

1.9 Duckweed on a pond

The growth of a population of duckweed (*Lemna minor*) on a pond, half of which was shaded by overhanging willow branches, was recorded at weekly intervals during late spring and early summer. Fig 1.14 shows the results of the survey.

Fig 1.14 Growth of a population of duckweed

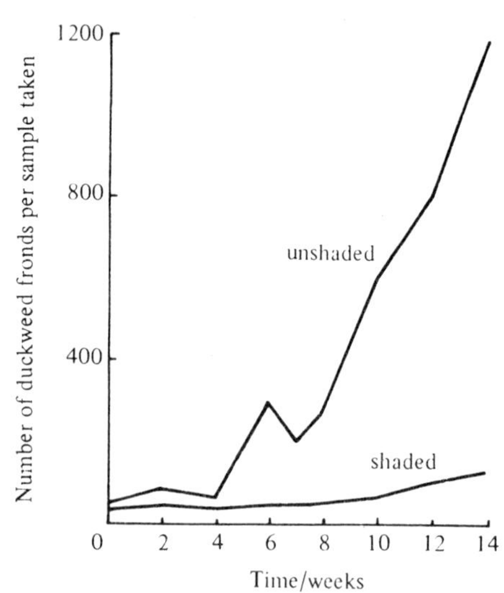

1 Suggest one hypothesis to account for the difference in numbers of duckweed in the unshaded and the shaded halves of the pond.

2 Suggest explanations for the dips in the curve for the unshaded side of the pond.

The effect of temperature was investigated by setting up ten beakers by the unshaded side of the pond, and ten beakers in a greenhouse at a temperature about 10 °C higher than outside. Each beaker, with its sides covered by black paper, contained pond water and 30 duckweed fronds.

3 What results would you expect from this investigation?

4 The actual results are shown in Fig 1.15. What do you think was limiting growth in the beakers by the pond?

Fig 1.15 Growth of duckweed in beakers

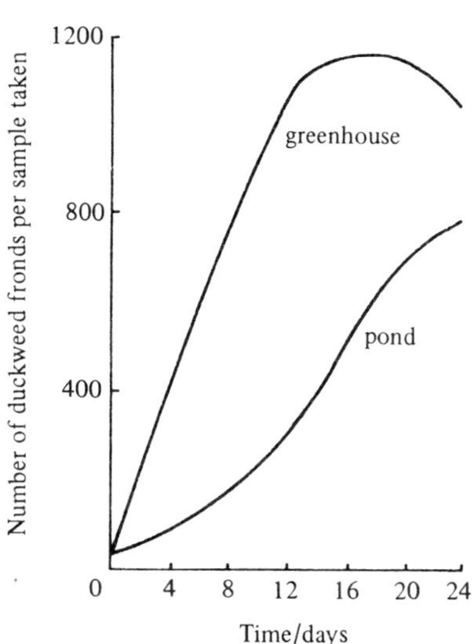

5 Suggest three reasons, any one of which could account for the duckweed in the greenhouse beakers entering the death phase.

6 Suggest a series of experiments which would enable you to decide whether the reasons you give in answer to **5** were in fact the cause of the population decline.

Fig 1.16 gives the results of a further investigation where

A 30 of the living greenhouse duckweed plants were left in the beaker in the same water, the remaining plants being removed
B 30 of the living greenhouse duckweed plants were put into a beaker containing fresh pond water
C 30 fresh duckweed plants from the pond were put into fresh pond water
D 30 fresh duckweed plants from the pond were put into a beaker containing water which had supported a greenhouse population, all old plants being removed.

7 **a** What do you think this investigation was set up to test?
b Which was the control experiment?

8 What would be your conclusions from the results given in Fig 1.16?

Fig 1.16 Results of the investigation into the growth of duckweed

A original water, original duckweed
B fresh water, original duckweed
C fresh water, fresh duckweed
D original water, fresh duckweed

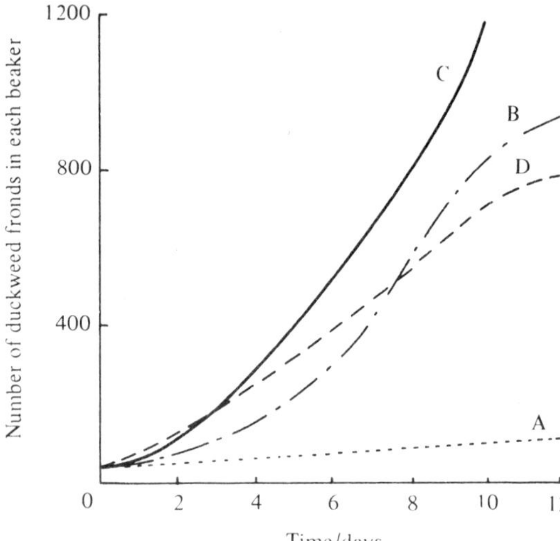

1.10 Barnacle populations

The barnacle, *Balanus balanoides*, is one of the two common barnacles on British coasts. It is one of several species of invertebrate which colonise ships' hulls and reduce the efficiency of movement, as well as colonizing many other man-made surfaces which may need expensive cleaning and maintenance from time to time.

The adult barnacle is fastened to the surface. It feeds by straining suspended food from the water, using feather-like limbs which protrude from the shell. The limbs sweep rhythmically through the water and strain off the suspended food. The larval form is not sessile. It is part of the plankton and swims for a few weeks before settling down.

An investigation was carried out to determine the effect of the density of initial settlement and the degree of water movement on egg productivity, which of course

Fig 1.17 Egg production by barnacles in three environments

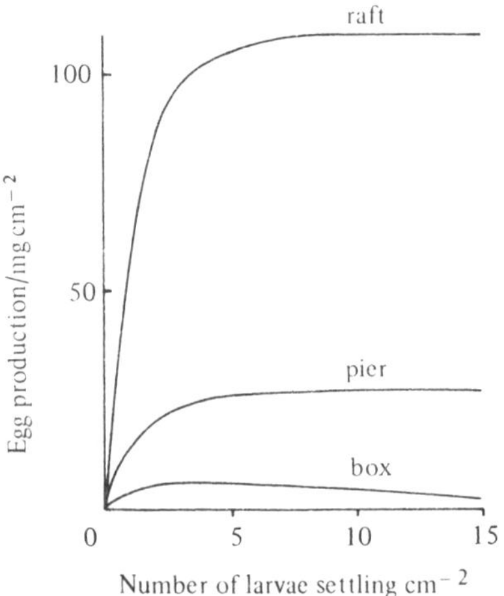

affected the final amount of fouling of ships' hulls. Barnacle larvae were allowed to settle on experimental panels at different densities. Three series of panels were used, each in a different situation. One was exposed from a raft which was kept in a rapidly flowing part of the sea. Another was fastened to a pier in the same area where it was in the sea for only half the time. The third was put in the same position but in a box which had small holes in it, thus greatly reducing water flow over the barnacles.

In each population, any other species which settled on the panel were picked off. The barnacle larvae settled in May and grew until December. All those which were able to breed would by then have contained fertilized eggs. Fig 1.17 gives the results.

1 State the relationship between the rate of water flow and egg production.

2 Suggest a hypothesis to account for this relationship, and an experiment which would test your hypothesis.

3 What kinds of man-made surfaces in sea water would barnacles colonize and, in doing so, foul?

4 What other factors could affect the growth of a barnacle population?

5 There will be intraspecific competition within the barnacle population. For what might the barnacles be competing?

2

INTERACTIONS

2.1 Predator-prey relationships

An organism's **biotic environment** is made up of all the other organisms with which it comes into regular contact. With some of these organisms it has relationships which may profoundly influence the distribution and size of its population. One such example is **predation**: one species feeding on another. The feeding relationship between the two species will determine the relative abundance of each, and may help to mould their evolutionary future.

This problem is about the balance between **predator** and **prey** populations. A population of predators is much smaller than the prey population from which it draws its food. Although there is usually some kind of balance between the two populations, this balance is rarely stable and may show periodic fluctuations.

1 **a** What will happen to predators if their prey becomes scarce?
b What could then happen to the prey population?
c If the predators are too efficient what will happen to both populations?
d If the predators are entirely removed what might happen to the prey population?

2 **a** Do you think that predation could be described as beneficial to the prey species?
b Do you think that predation is also beneficial to the predator species, apart of course from providing it with food?

The figures show predator–prey relationships in two different communities. Each one is explained.

Fig 2.1 Populations of prey and predator on an orange

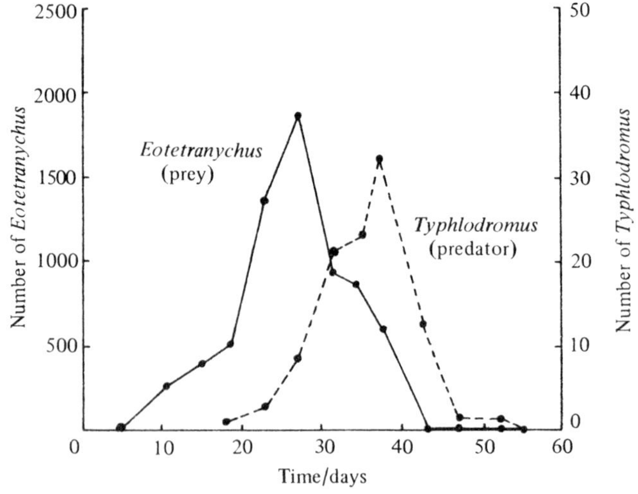

Fig 2.1: Two species of mite, one of which (*Eotetranychus sexmaculatus*) feeds on oranges, and the other (*Typhlodromus occidentalis*) which feeds on the first, were kept in the laboratory. Early experiments either allowed the predators to move about easily from orange to orange or put the predators onto a single, prey-infested orange. The results were as shown in Fig 2.1.

Fig 2.2: Later, in a more complex situation, the predators were slowed down by streaking Vaseline between groups of oranges. The predators could not cross over the Vaseline. The prey were helped to disperse by being given 'launching posts' from which they (but not the predators) could float on strands of silk to other areas. Fig 2.2 gives the results.

Fig 2.2 Populations of prey and predator on several oranges

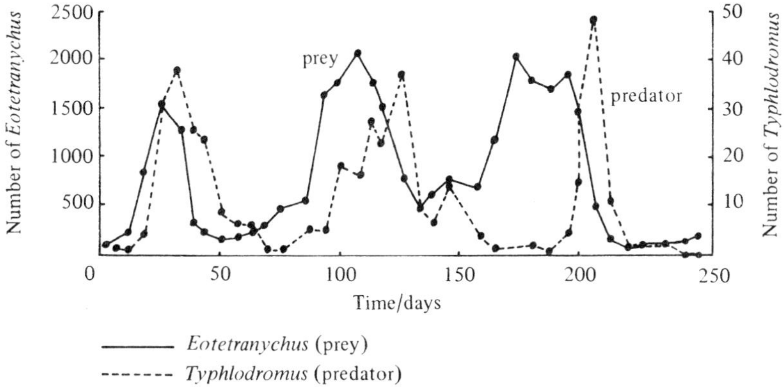

3 Comment on each of these situations with regard to the levels of predator and prey populations, and the stability of the populations related to the complexity of the environment.

Fig 2.3 Cycles of abundance of snowshoe hare and lynx

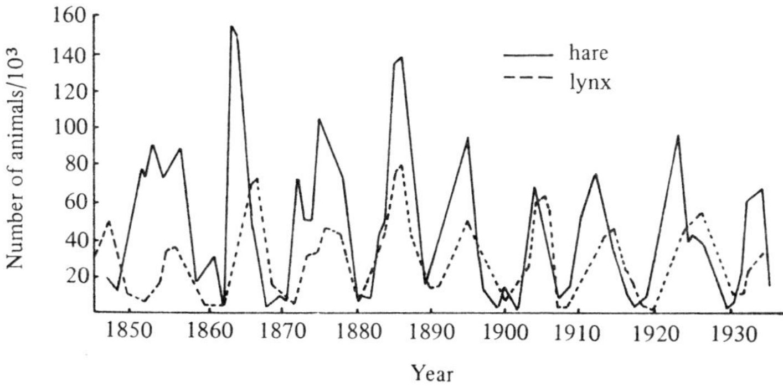

A classic example of predator–prey relationships is given by the numbers of the Canadian lynx (*Lynx canadensis*) and the snowshoe hare (*Lepus americanus*) which live in the cold northern parts of Canada. Fig 2.3 shows the numbers of each species, estimated from four sources: the records of the number of skins received by the fur trading company, statements in the literature, questionnaires to trappers and other people, and field work which involved trapping, counts of hares seen and counts of droppings. Numbers of the hare per square mile ranged from one when they were scarce to 1000 when abundant, and up to 3400 as the highest number. Lynx abundance is governed by that of hares, which form 80% to 90% of their diet.

4 a How does Fig 2.3 support the statement above about lynx abundance?

b The cycles of hare abundance are fairly regular, the average length of a cycle from one peak to the next is 9.6 years. The fluctuations are huge: an increase from 15 000 to 135 000 in three years in one cycle and a fall from 155 000 to below 5000 in four years in another. Suggest a hypothesis which could explain these cycles, and the enormous fluctuations.

Food for the hares is abundant in the summer but low in winter. Nearly all the recorded dead hares and decreases in number have been found in the summer, and there seems to be a wholesale dying off during a population decline. The hare population decreases so far that the available food is hardly utilized for a number of years.

c Does this extra information cause you to think again about your answer to **b**? If so, suggest a further hypothesis to account for the fluctuations. Is your suggested cause of the population crash a density dependent factor or a density independent factor (see Problem 2.4)?

d Over one hundred years, the hare abundance cycles varied in length from eight to eleven years. Why do you think they were so regular and approximately the same length?

People have become the most important biotic factor in the world. Our interference with the natural order of things can bring about disaster. A classic story is that of the deer population on the Kaibab Plateau in Arizona. Prior to 1906 the plateau supported a herd of about 4000 deer, as well as some domestic grazing animals. The population of deer was kept to about that number by the pumas, wolves and coyotes which preyed on the deer. When the area was declared a wild-life refuge in 1906 the domestic livestock were removed and the 'gentle, defenceless' deer were protected by the systematic extermination of their predators.

Fig 2.4 The deer population on the Kaibab Plateau

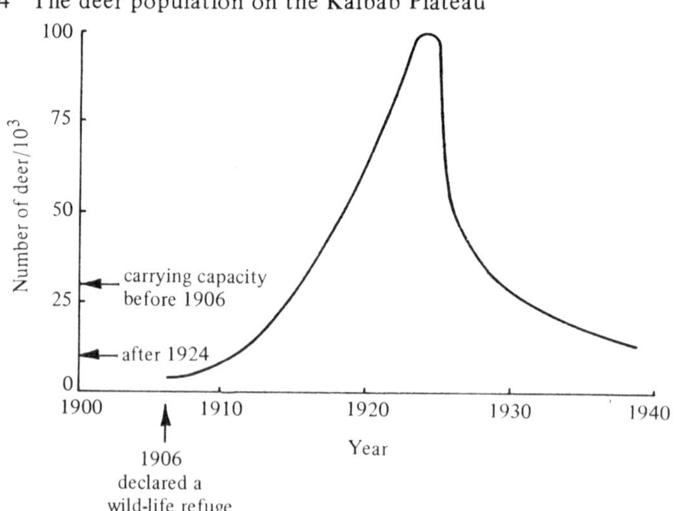

After the plateau had become a wild-life refuge, 3000 coyotes, 600 pumas and 11 wolves were killed. The result was that the deer population increased enormously, possibly to as many as 100 000 in 1924. It was certainly well above the 30 000 which the vegetation on the plateau could theoretically support, called its **carrying capacity**. Fig 2.4 shows this story graphically.

5 Finish the story of the deer on the Kaibab Plateau by studying the graph. Say what you think would have happened if the natural populations had been left undisturbed.

Man may upset the natural balance of predator and prey in other ways. Consider the situation where a careless or ill-judged use of insecticide in an ecosystem has destroyed not only most of the population of a plant-eating insect pest, but also all of its predators.

6 a What problem could then arise with the pest, and how might it be overcome?

b On the other hand, suppose the insect pest is completely removed by an insecticide. Its predator is left with not enough food to survive and dies off. What could happen to the populations of other pest species, too small to support the predator on their own, which the predator had previously helped to control?

7 In summary, which of the following would result in the most stable situation?

a A simple environment with one prey species for each predator.
b A complex environment with one prey species for each predator.
c A simple environment with many prey species for each predator.
d A complex environment with many prey species for each predator.

2.2 Coloration and mimicry

One way in which animals can escape the attention of their predators is seen in those, such as the peppered moth, which have **cryptic coloration**. The body colour of such animals is similar to that of their normal habitat and enables them to fade into the background. Some insects have a shape as well as colour which matches the background. They look like leaves or twigs. They must, of course, also behave in a suitable way: a leaf insect resting on a tree trunk, or a stick insect running along a branch, would quickly attract attention.

Disruptive coloration is another mechanism which helps an organism to escape predation. The animal is marked with a disruptive pattern of bands, patches or spots which hides the body outline. A zebra is one example, and another is shown in Fig 2.5. The predator sees only the pattern, not the animal.

Some cryptically coloured prey animals have another strategy. They show a bright flash of colour when they move. The predator searches where the prey was seen to jump from, while the prey safely fades into the background a little further on. Other species, such as some moths, frighten predators when disturbed by the sudden appearance of an eye pattern on the hind wings (Fig 2.5), or have other types of frightening behaviour.

Other organisms have a different method of escaping predators. They have a noxious taste or some other effect on the predator when eaten. Such distasteful or unpalatable species often have **warning coloration**: a striking colour pattern which serves as a warning to predators who have once tried the species. Predators learn very quickly to avoid the black and yellow stripes of a wasp. The first part of this problem concerns such an unpalatable species.

Fig 2.5 Escape from predators

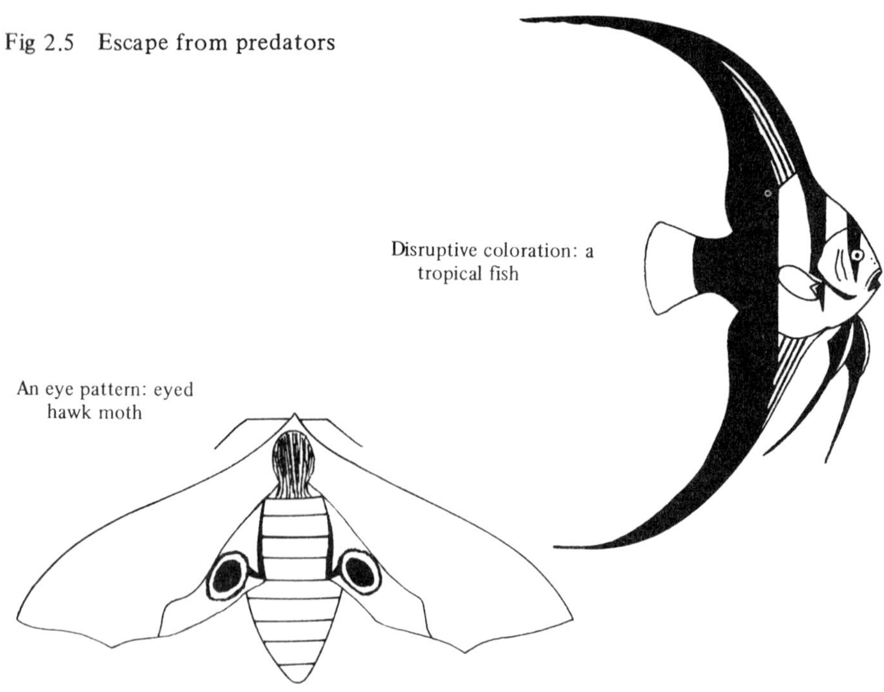

Fig 2.6 The butterfly, *Heliconius erato*, with altered wing pattern and wing damage

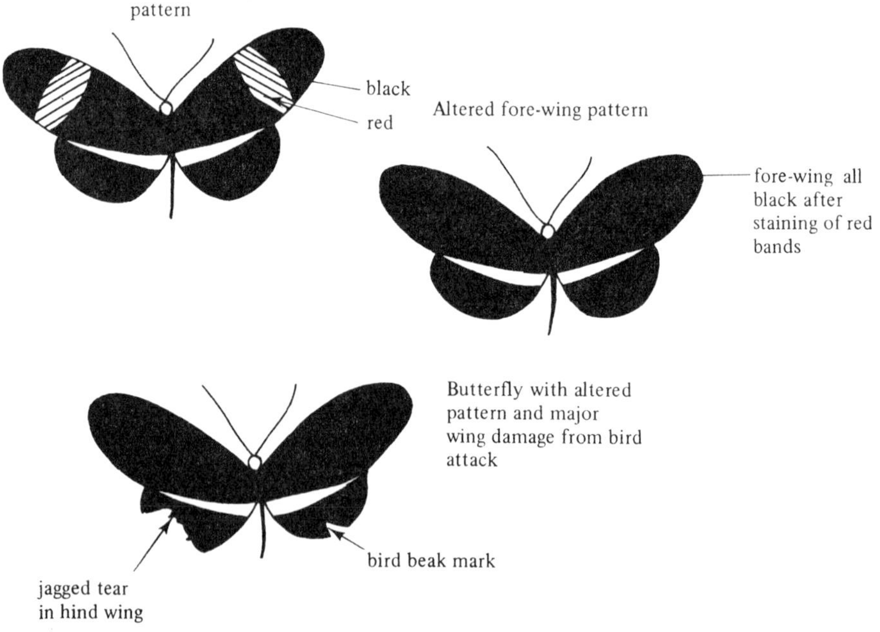

In Costa Rica, an unpalatable species of butterfly, *Heliconius erato*, was investigated. The butterfly has a striking red and black pattern on its fore-wings. The experimental design involved altering the red and black pattern. This was done by staining black the red band so that the whole of the fore-wing appeared black. The butterflies returned to the same place each night and roosted together. Two measures of survival were made each night: whether or not there was major wing damage (Fig 2.6) which would probably have been caused by a predator; and the number of nights a butterfly returned to the roost. A butterfly was assumed dead when it failed to return. The results are given in Table 2.1.

Table 2.1 The effect of altering the wing pattern on survival of a butterfly

	Butterflies with major wing damage/%	Number nights returned
Altered	42	32
Unaltered	12	52

1 What conclusion does it seem can be drawn from these results?

2 Before you accept the results you will want to know other facts which have not been stated here. For example, what have you assumed about

a the effect of the stain used
b the effect of observation and handling on butterfly behaviour?

A third group of the same species was treated as follows. The already black wing tips of the fore-wing were stained with the black stain. This did not affect the colour pattern. Results for this group were very similar to those for the unaltered butterflies.

3 What was the point of this treatment, and does it go some way to meeting any point you mentioned in your answer to 2?

Another way in which some palatable species avoid their predators is by **mimicry** of an unpalatable species. The palatable species closely resembles the unpalatable one. One of many experiments on mimicry involved two species of insect which could be taken by toads. It showed that toads fed live bees which could sting thereafter avoided not only bees but also a palatable drone fly which looks like a bee. If the toads were fed dead bees from which the sting had been removed, they also ate the fly mimics without any hesitation.

4 How does this experiment show the principle of mimicry?

Another experiment, which was done in Trinidad, involved a palatable Saturniid moth (*Hyalophora promethea*) and a distasteful butterfly, *Parides*. The predators were birds. Survival rates were estimated by recapture methods. The moths were released and later recaptured by attracting them to caged female moths.

In all three experiments described below, the control moths were painted black over their black wings so they received a similar amount of handling and paint as the experimental moths did. The experimental moths were painted differently in the three experiments. In experiment 1 they were painted conspicuously but unlike any butterfly or moth in Trinidad. In experiments 2 and 3 they were marked red, black and green to resemble the distasteful butterfly. The difference between experiment 2 and experiment 3 lay in the numbers released at any one time. In experiment 2 the moths were released altogether in one place, in experiment 3 they were released at different times and at different places. The results are given in Table 2.2.

Table 2.2 The effect of altering the colour of a moth on its survival

	Expt. 1 (Experimental moths conspicuous and unlike any other)		Expt. 2 (Experimental moths resemble distasteful *Parides*)		Expt. 3	
			Many released same time and place		Released over longer time	
	Control	Experiment	Control	Experiment	Control	Experiment
Numbers released	42	43	414	414	194	199
Numbers recaptured	16 (38%)	8 (19%)	129 (31%)	95 (23%)	37 (19%)	53 (27%)

5 What conclusions may be drawn from these three experiments with moths? Comment particularly on the experimental design of experiments 2 and 3.

Batesian mimicry is where an uncommon but palatable species (the mimic) resembles a common distasteful species (the model). Predators learn to avoid the model and later avoid the mimic also. **Müllerian mimicry** is where two or more unpalatable species resemble each other. Predators have only one pattern to learn, and the unpleasant experience of eating the unpalatable prey is spread over both species.

6 Are the results in Table 2.2 consistent with the Batesian hypothesis?

Occasionally a species seems to mimic itself: **automimicry**. The caterpillars of the monarch butterfly feed on milkweeds, some species of which produce a chemical (glycoside) which makes the butterfly unpalatable and causes the bird predator to show all the symptoms of poisoning, including vomiting. Caterpillars which feed in an area where both glycoside and non-glycoside milkweeds grow produce butterflies of varying degrees of palatability, depending on the caterpillar's diet.

7 What two advantages are there to the insect of being able to resist and store poisonous chemicals from its food?

2.3 Interspecific competition

Competition is the name given to the interaction between populations in which each population adversely affects the other in the finding or utilization of food, nutrients, living space, light or some other common need.

Interspecific competition is any interaction between two or more species populations which adversely affects their growth and survival. If, for example, the growth curves of two populations are both steeper when they are separate than when they are interacting, competition of some sort is operating. Competition can give an equilibrium between two species, or it can result in one species replacing another or utilizing another food source or otherwise behaving differently.

In the main, organisms with similar needs do not occur in the same place. If they do they may adapt and occupy slightly different niches by, for example, being active at different times or diversifying their food sources. A **niche** is the position or status of an organism within the community resulting from its adaptations, physiology and behaviour, including its energy source, period of activity and so on. An organism's niche depends not only on where it lives but what it does there. The American ecologist E. P. Odum calls the niche the organism's profession. Competition occurs wherever two niches overlap to any extent.

It is difficult to obtain direct evidence of competition from the wild, partly because of the difficulty of counting numbers and measuring mortality in natural populations. A lot of work therefore has been done with populations in the laboratory. Some of the best documented experiments of population biology were done many years ago by the Russian biologist G. F. Gause, and form classic cases. Some of Gause's experiments, in 1934, involved two species of *Paramecium*.

Paramecium caudatum and *P. aurelia*, when they were cultured separately in the laboratory in uniform conditions and fed on bacteria, showed the usual curve of population growth. At the top of the curve they maintained a more or less constant population (Fig 2.7). There were fewer individuals of *P. caudatum*, the larger of the two species, than of *P. aurelia* in each final population. When a third culture was started, with equal numbers of both species, the population of each developed as shown in Fig 2.8.

Instead of plotting the number of each species, Gause used volume. He counted the volume of the larger *P. caudatum* as equal to one. The smaller *P. aurelia* then had a volume of 0.39 of that of *P. caudatum*. The 'volume' of *P. caudatum* is thus the same as the number of individuals. The number of individuals of *P. aurelia*, however, is multiplied by 0.39 to give their 'volume'.

Fig 2.7 (left) Growth of populations of two species of *Paramecium* when each is cultured separately
Fig 2.8 (right) Growth of populations of two species of *Paramecium* when they are cultured together

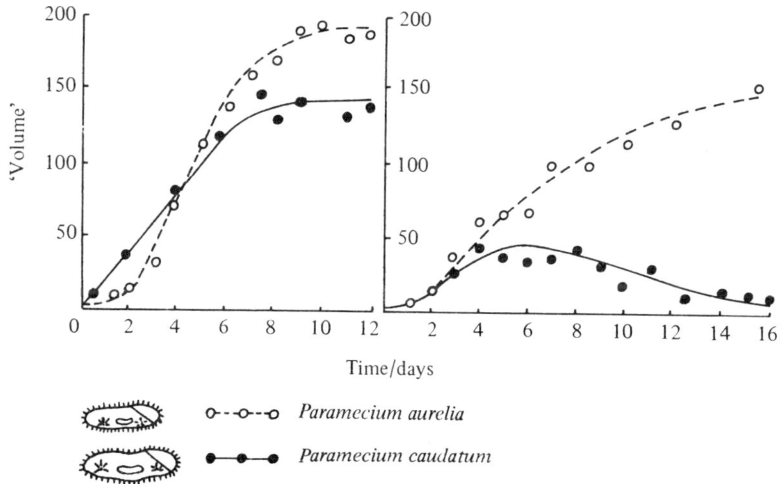

1 Describe and interpret what is happening to each species when they are cultured together.

2 We are supposing that these two closely related species are competing for some resource in the culture. What resources could they be competing for?

3 Apart from competition, what other explanations could be suggested to account for the success of one species at the expense of the other?

4 Assume that there is competition for food. What factors may be important in the rise of one species and the fall of the other? In other words, what causes one species of animal to be successful or to be eliminated when two species compete for the same food supply?

5 Consider the theoretical example of a culture flask which will support a maximum of one hundred *Paramecium* individuals. The culture is started with equal numbers (50) of two species. Suppose that one species increases by about three per cent per generation, while the total number of individuals stays at 100. How many generations will it take for the unsuccessful species to disappear?

6 Instead of *Paramecium* in a flask reproducing either sexually or asexually, consider a decreasing population of a vertebrate species in competition for food with another more successful species. What other factors would hasten the decline of the unsuccessful species?

2.4 Competitive exclusion

A series of experiments on *Paramecium* was discussed in Problem 2.3. Another series of experiments on interspecific competition, this time by Crombie in 1946 and Park in 1948, 1954 and 1962, used the flour beetles, *Tribolium castaneum* and *Tribolium confusum*. They are useful laboratory animals as the whole life cycle can take place in a jar of flour which is both food and habitat for the animals. These experiments also demonstrated **competitive exclusion**: the exclusion of one species by competition, already seen in Problem 2.3. Both species could not survive together in the simple ecosystem of a bottle of flour. Which species survived depended on the conditions inside the bottle.

When each species was grown separately, each could survive in each set of conditions tested, although both actually did better in warm, moist conditions. When they were cultured together, however, typical results were as shown in Table 2.3.

Table 2.3 Interspecific competition between populations of two species of *Tribolium*

Climate	Replicate experiments in which each species remained alone after the other had died out/%	
	T. castaneum remains	*T. confusum* remains
Hot–wet	100	0
Hot–dry	10	90
Warm–wet	86	14
Warm–dry	13	87
Cold–wet	29	71
Cold–dry	0	100

1 Which conditions favour each species?

2 The more tolerant a species is and the wider the limits of its existence, the

more chance it has of surviving in competition. How do the results in Table 2.3 demonstrate this principle?

3 If the cultures were infected with a sporozoan parasite, *T. confusum* usually won in competition. What conclusion about life in the wild as compared with life in a bottle can you draw from this?

When *Tribolium* is put in culture with another (smaller) beetle, *Oryzaephilus*, the *Tribolium* destroys the other by killing the young stages (Fig 2.9). If, however, fine glass tubes are placed in the flour, the young stages of *Oryzaephilus* can escape into them. Both populations survive and co-exist together.

Fig 2.9 Populations of two species of beetle in flour, with and without glass tubing

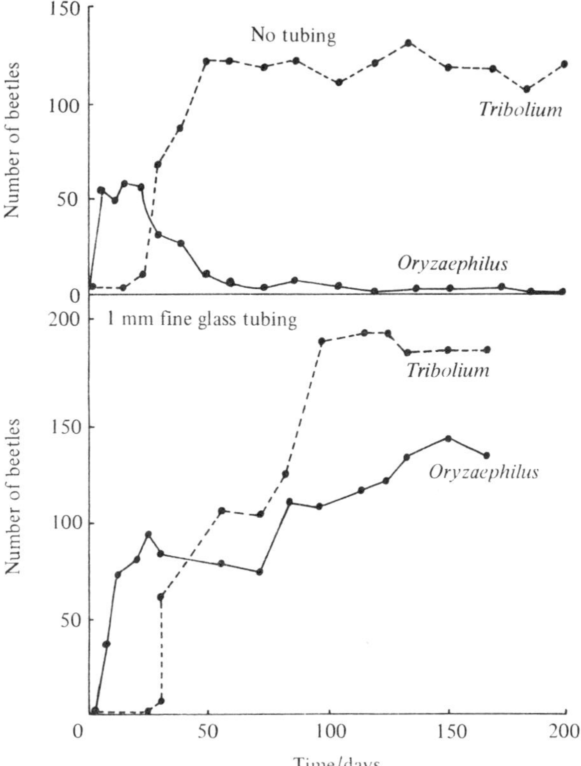

4 Remember that the beetles are in an enclosed, simple and homogeneous environment. What principle do you think this experiment demonstrates? (Refer to Problem 2.1: Fig 2.2 and questions 3 and 7.)

You may meet the terms **density dependent** and **density independent** in your reading, usually referring to factors which affect the death rate within a population. A density independent effect or factor is one which affects the organism as an individual. If a density independent effect such as an unusually high or low temperature is going to kill, it does so regardless of whether the population is dense or sparse.

A density dependent factor however has more effect on mortality within a dense population than within a sparse one. In this problem, the parasite referred to in question 3 is an example of a density dependent mortality factor. Very cold conditions killing off both populations would be a density independent factor.

5 List any two other density dependent or density independent factors which could affect mortality within an animal population.

2.5 Specialization through competition

Barnacles are animals covered by a shell. They live fastened to coastal rocks in the area which is exposed at low tide, called the intertidal zone. Once attached to a rock barnacles are fixed to the spot until they die. They feed on zooplankton when they are covered by the sea.

In Scotland, one species of barnacle, *Balanus balanoides*, occurs in the low part of the intertidal zone. Another species, *Chthamalus stellatus*, which is of a similar size to *Balanus*, occurs higher up on the rocks (Fig 2.10). Young *Chthamalus* often attach themselves in the lower shore, but no adults are found there if *Balanus* is present in numbers. If there is no *Balanus*, *Chthamalus* can easily live in the lower shore. On the other hand, even if there are no *Chthamalus*, *Balanus* cannot establish itself in the upper parts of the shore.

1 Why do you think *Chthamalus* is restricted to the upper shore?

Fig 2.10 Distribution of two species of barnacle on a rocky shore

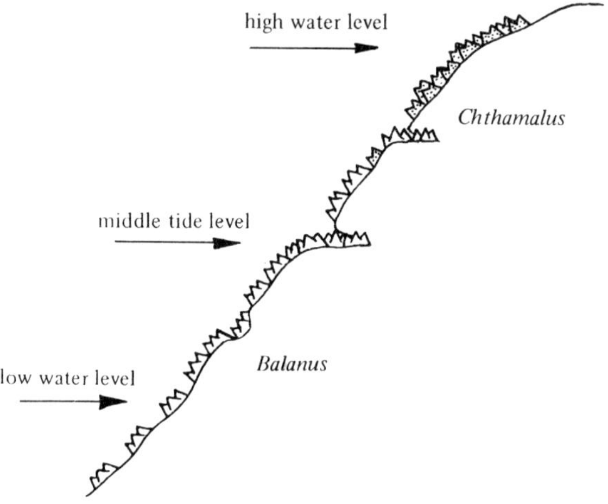

2 If this is an example of competition, what do you think the two species are competing for?

3 Design an experiment to determine whether the two species are competing for the resource you named in answer to 2.

4 How does this example show that the more tolerant a species is, and the wider the limits of its existence, the more chance it has of survival?

2.6 Competition and evolution

Evidence from natural populations shows that competing populations can co-exist if the limits of their tolerance are not too narrow and there is enough variation in the limiting resource to allow each to specialize within that variation. Problems 2.4 and 2.5 gave examples. Even the absence of competition between two similar species living side by side may only indicate that competition in the past has brought about the development of the now distinguishing behaviour or other characteristics.

Fig 2.11 Diagram to show overlap between two species competing for a resource

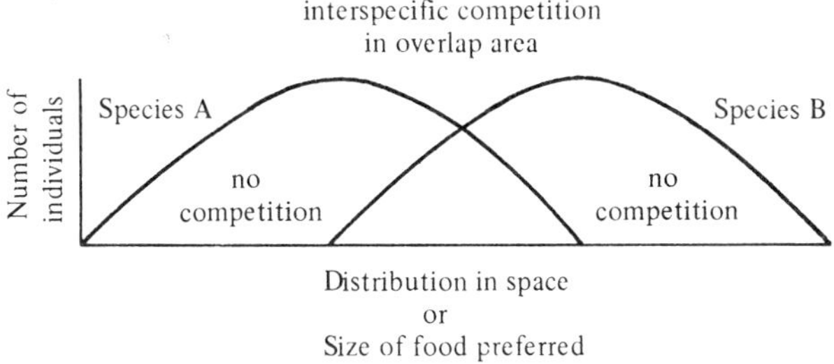

Consider hypothetical situations, referring to Fig 2.11. In one situation, the distribution or range of two similar species overlaps considerably. The pressure on the same **resource** which they both need, say food or a habitat, is heavy where they overlap. In this overlap area there is intense competition between them for the resource. At the ends of the distribution there is very little, if any, interspecific competition, and each species flourishes alone.

In another situation, two similar species live in the same area, overlapping considerably in their use of a resource. Perhaps they both feed on insects. One prefers middle to large size insects, the other prefers middle to small. Again, in the overlap area of middle-sized insects there is intense competition for the food resource. At either end of the food supply (small insects and large) there is no competition.

1 What will happen to many of those individuals of both species which live in the overlapping area or which prefer the middle size of insect?

2 How could this affect the genetic constitution of the two populations which are left?

3 Draw graphs of the two populations shown in Fig 2.11 after the process which you described in answer to question **1**.

It is of course difficult to show this process happening over a long period of **time**. One can, however, consider existing situations in **space** where the populations of two similar species overlap for a part of the total range of each.

4 In such a case, where would you expect the differences between the two species to be greatest: in the overlapping area of their ranges, or in the areas where they do not overlap?

Figs 2.12 and 2.13 illustrate two examples of overlap. Fig 2.12 gives the range and length of beak of the western rock nuthatch (*Sitta neumayer*) and the eastern rock nuthatch (*S. tephronota*). They are small birds, very similar to each other, which probe for insects on the ground. The range of the western rock nuthatch is from Yugoslavia, across Greece and Turkey to Iran. The eastern rock nuthatch is found from Iran and across Afghanistan.

Fig 2.13 gives frequency distributions of beak sizes of two species of Darwin's finches (*Geospiza fuliginosa* and *G. fortis*) on eight of the isolated Galapagos islands. Both species feed on seeds. Beak size is related to the size of the seeds on which each species feeds. Similar types of seed occur on all eight islands.

Fig 2.12 Geographical range and beak length of two species of rock nuthatch

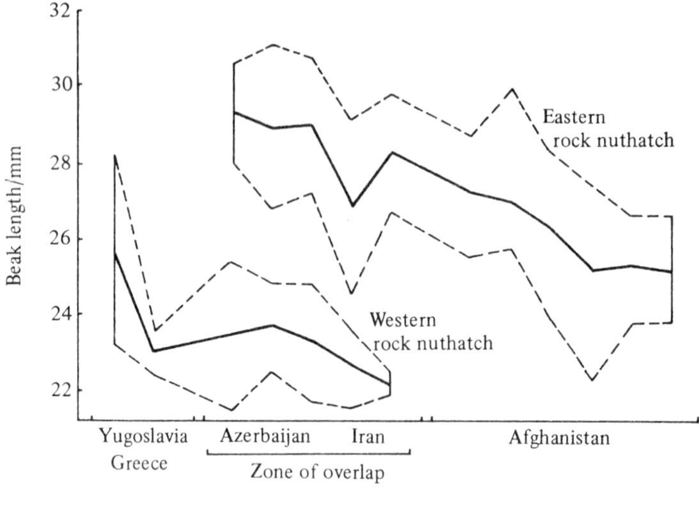

Fig 2.13 Frequency distribution of beak depth of two species of ground finch (*Geospiza*) on eight of the Galapagos Islands

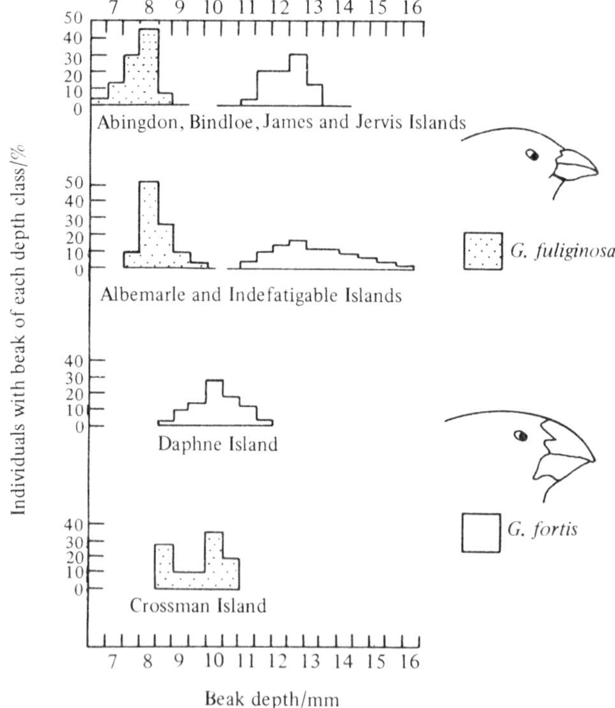

5 Competition can lead to divergence and hence evolution. Say how the two examples above could be used to support this statement.

2.7 Interspecific competition in plants

The perennial grass, *Agropyron spicatum*, used to be the dominant plant species in the dry region of the north-west United States. After the introduction of European agricultural practices the composition of the grassland changed. Annual grasses, of which the most frequent was *Bromus tectorum*, invaded grassland previously dominated by the more useful grazing grass *Agropyron* and became dominant in its place. Many experiments were done to try and understand the nature of the competitive advantage possessed by the *Bromus* in an effort to get the grassland back to *Agropyron*. The results of two investigations follow.

When grown in dense, moderate and sparse populations of *Bromus*, the average survival of *Agropyron* was 39%, 69% and 86% respectively of those planted. The average root lengths of the two species when grown separately and then together at different densities are given in Table 2.4.

Table 2.4 Root lengths of two species of grass when grown separately and together/cm

	Agropyron	*Bromus*
Grown separately	64.4	82.8
in the ratio of 4 *Agropyron*:1 *Bromus*	56.9	85.4
1 *Agropyron*:1 *Bromus*	47.2	91.1
1 *Agropyron*:4 *Bromus*	41.0	82.1

1 What evidence is there from the data of competition between the two species?

2 Consider the structure of a plant and a typical plant life cycle. What features of structure or life cycle might contribute to the competitive advantage of one species over another? In other words, what features might cause one species of plant to be successful or to be eliminated when two similar species live side by side?

A relatively large number of species can live together in a small area of vegetation which looks fairly uniform. In a rich grassland, for example, there could be up to fifty species of plant in an area of about twenty-five square metres. The variation in water supply, nutrients and so on over the area could be slight. However, species may compete differently at different times of their life cycle, or at different times of the year with differing climate and light periods. One species does well at one time, another at another time, and a **competitive balance** exists between all the species which survive there.

3 For which resources in any one habitat might two species of plant growing side by side compete, or which may they exploit differently?

2.8 Competition between wheat and weeds

In many cases we try to eliminate interspecific competition between a crop and weeds so that the yield of the plant being grown for food is not reduced. Sometimes careful decisions, based on the estimated selling price of the crop, the cost of weeding and the various other operations needed to obtain a good crop, have to be made. It may pay to remove the weeds, or it may not. The effect of the weeds may perhaps be overcome by fertilizer or water which could be easier or cheaper to apply than weeding the crop would be.

Research evidence can help a crop grower to decide what to do, or influence the advice given to crop growers, so that the experience and knowledge of a farmer is reinforced by experiments under controlled conditions.

Such a series of experiments was done in Australia using wheat and a weed, *Emex australis* (three-cornered Jack), together with the effect of a nitrogenous fertilizer. In the experiment three levels of nitrogen were used: no nitrogen (N_0), 6.3 g fertilizer m^{-2} (N_1), and eight times that, 50 g m^{-2} (N_8). They were used together with four levels of weeding where the crop was weed-free all the time (W_0) or where weeds were removed on one of three occasions: in mid-September (W_1), mid-October (W_2), mid-November (W_3). Harvest date was December 18th.

Table 2.5 gives the yield of grain and straw for all treatments. Straw is all the aerial parts of the plant except the grain.

Table 2.5 Yield of straw and grain of wheat/g m^{-2}

	Level of nitrogen application					
	N_0		N_1		N_8	
Level of weeding	Straw	Grain	Straw	Grain	Straw	Grain
W_0	478	245	549	284	622	324
W_1	343	213	402	223	596	287
W_2	336	196	358	203	488	248
W_3	289	140	326	169	505	213

1 What is the effect on the yield of wheat of

a time of weed removal, b added nitrogen?

2 The data in Table 2.5 relate to the yield of both straw and grain. In fact, are both parts of the plant useful?

When only the grain was considered and the number of grain-bearing stalks on the plant counted, the results given in Table 2.6 were obtained.

Table 2.6 Number of grain-bearing stalks per plant

	Level of nitrogen application		
Level of weeding	N_0	N_1	N_8
W_0	4.1	4.7	5.3
W_1	3.0	4.2	5.3
W_2	2.7	3.6	4.3
W_3	2.3	3.3	3.7

3 a In which plot, W_0 to W_3, was the effect of nitrogen most pronounced?

b What resource do you think the weed and crop plants could be competing for? Give the reasons for your answer.

4 a Is competition for nitrogen the only factor involved? Give reasons for your answer.
b If some other factor is involved in this competitive relationship what do you think it might be?

This problem has indicated how even a short infestation of weeds can reduce the yield of a crop, and how even a small amount of added fertilizer can increase it.

5 Suppose you were yourself having to take a decision about whether or no, and if so how, to treat this crop. What facts would you need to know before you initiated some action?

2.9 Grazing

A **pasture** is an area of land, or a field in which is growing a mixed community of herbaceous plants, dominated by grasses. Such fields are grazed by herbivorous farm animals.

1 Grasses are particularly suitable plants for grazing. Why?

Pasture for farm animals has to be managed so that the right species continue to grow there, and to grow well. In addition, pastures are semi-natural places (like hedgerows) where a lot of species of wild flowers can live.

Different farm animals graze in different ways. All of them choose leaf in preference to stem, and green young material in preference to dry, old material. Sheep, however, graze the plants quite close to the ground, as compared with cows which graze about ten centimetres above the ground. Sheep are more selective than cows in their choice of mouthfuls. The composition of the pasture is therefore altered in differing ways by these two grazers.

Some grazing, for example by rabbits, can be destructive to pastureland. Rabbits nibble selectively and close to the ground, and also destroy taller plants of species which they do not eat. Unlike sheep and cows, rabbits burrow into the earth and create bare areas around the burrows which re-establish only slowly. Since the disease myxomatosis killed many rabbits in 1954–5 they have not been such a pest as before.

Grazing, of course, affects some species of plant more than others. The grasses can stand up to grazing very well, but other species can be severely affected, either immediately or by destroying their competitive edge over their neighbours.

2 Many investigations have been done into the effect grazing has on the species composition of a pasture. A lot of these investigations use a fairly simple experimental design. What do you think would be a suitable experimental method to use? What control experiment would you have? Predict the results from the experimental method you suggest.

3 Do you think that grazing leads to a larger or smaller number of species in a pasture? Give the reasons for your answer.

2.10 Intraspecific competition in plants

Sometimes a population of a single species is so dense that the individuals cannot achieve their maximum growth. Some one or more resources are limiting. The term **intraspecific competition** is given to competition between individuals of the same species. Density effects in plants can be demonstrated and measured experimentally.

Experiments with clover compared the effects of density of planting on the final yield (dry mass) of plant material which was harvested. All densities of clover were supplied with plenty of water and nutrients. Fig 2.14 shows the yield per unit area when the clover was planted at the densities shown along the horizontal axis. The plants were harvested at three stages of growth: when they were at the seedling stage, at 131 days after germination, and at 181 days.

Fig 2.14 Yield per unit area of clover at different densities and at different times

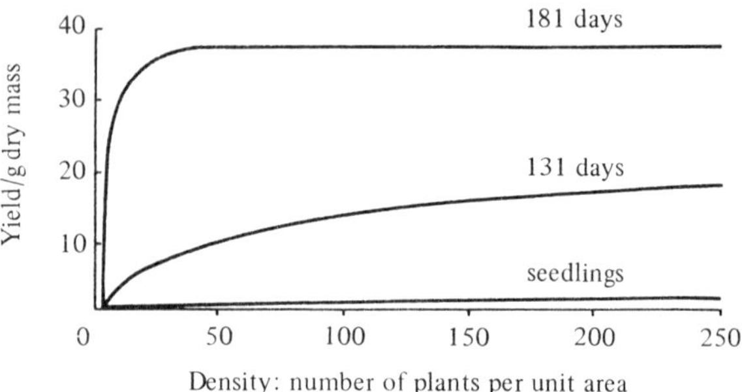

1 Say how these curves show competition between the clover plants.

2 Which resource do you think they are competing for and what would be the mechanism of the competition?

A second example of intraspecific competition from the plant kingdom is from an investigation with shepherd's purse. The results are shown in Fig 2.15.

Fig 2.15 Number of seeds produced by plants of shepherd's purse sown at different densities

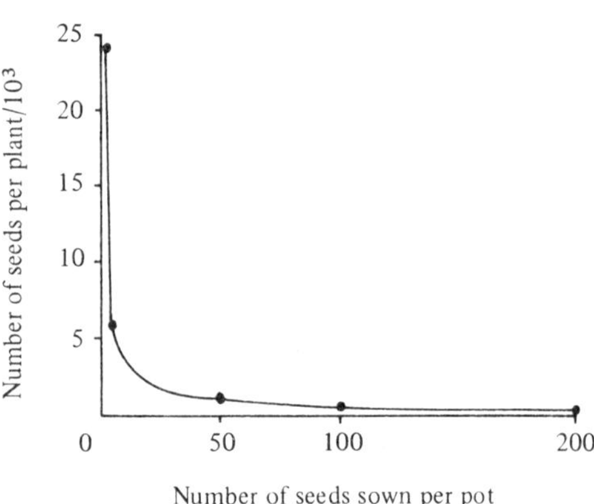

3 From these results, say briefly what you think the investigation involved and what conclusions may be drawn.

4 If one were considering the density at which to plant a crop, what points do you think should be considered? In other words, what would be the advantages and disadvantages of sowing at a high density?

5 In Problem 2.6 we saw that interspecific competition can encourage the evolutionary divergence of populations. Do you think it is advantageous for individuals within a population to vary from each other?

2.11 Inhibition

The effect of one species on another need not be limited to the indirect effects of competition, but could be due to some other factor. For example, there may be

Fig 2.16 Diagram to show the experimental design used to investigate the effect of a weed on the yield of flax

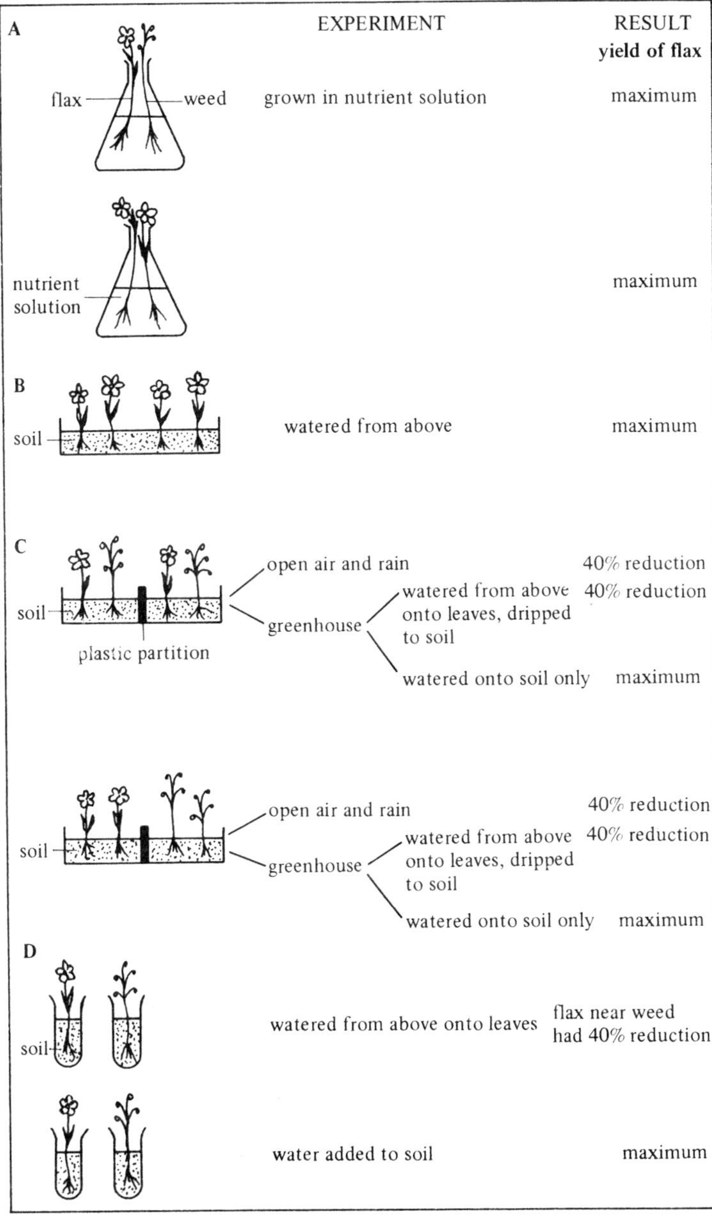

inhibition of one organism by another. An organism may release a chemical substance which inhibits the growth of another organism, or even kills it. The most well known example is probably the inhibition of micro-organisms by other micro-organisms. Alexander Fleming, in 1929, described the effect on bacterial growth of the mould *Penicillium notatum*. The inhibitor was a water-soluble compound which was later isolated as **penicillin** and formed the first of the **antibiotics**. Many other micro-organisms produce antibiotics used in medicine.

This problem gives two examples of inhibition.

Tadpoles hatch from frog (*Rana pipiens*) spawn and begin to grow. Individual tadpoles grow at different rates. In a large bowl, those which hatch first and grow quickly continue to grow. Those which hatch later grow more slowly. If the smaller, slower-growing tadpoles are left with the larger ones, the smaller ones stop growing and eventually die. If they are removed and put into water on their own they can start to grow rapidly.

1 Apart from the production of an inhibitor by the larger tadpoles, suggest two other hypotheses which could explain the results.

In fact, the effect of large tadpoles on small is due to water-carried particles produced by the larger tadpoles.

2 Design an experiment to demonstrate this. Say what results you would expect.

Flax is a crop plant grown for its fibres. *Camelina alyssum* is a weed. The presence of the weed in a crop of flax can reduce yield considerably. In experimental pots yield was reduced by over eighty per cent.

A series of experiments was set up to try and discover the nature of the interaction between the two species. The diagrams in Fig 2.16 summarize the experiments and the results.

3 How does the experimental design rule out

a competition between whole plants of flax and the weed
b competition between the roots of the two species?

4 Which experiments were included to determine if there was inhibition by the roots of the weed?

5 What was the purpose of experiment B?

6 It was found that an inhibitor was involved. What conclusion can you draw from the results about the source of the inhibitor? Explain your answer.

2.12 Flies and bacteria

Amongst the several species of two-winged flies found in many houses are some which are commonly suspected of carrying pathogenic bacteria.

1 Design an investigation to test the hypothesis that flies commonly carry bacteria which might be pathogenic.

2 Design a further investigation to see whether individual flies are effective distributors of bacteria from a source of infection, such as infected food.

3 It is said that some species of domestic fly carry a wide variety of micro-organisms. What method would you use to discover a minimum estimate of the number of species of micro-organisms carried by any one fly?

Flies which emerge from the pupal case are relatively free of bacteria and might even be absolutely free. It has been found, however, that fly larvae (maggots) do not develop properly in a bacteria-free medium and it was deduced that the maggots need a substance produced by the bacteria.

4 Suppose a substance is produced by the bacteria. How would you find out whether it was necessary for the bacteria to be alive if they are to help the fly larvae to develop?

5 If you did find that live bacteria were required, could you say whether the relationship between larva and bacteria was a case of symbiosis, co-operation, parasitism or predation?

3

ECOLOGICAL ENERGETICS

Fig 3.1 Energy flow through a community

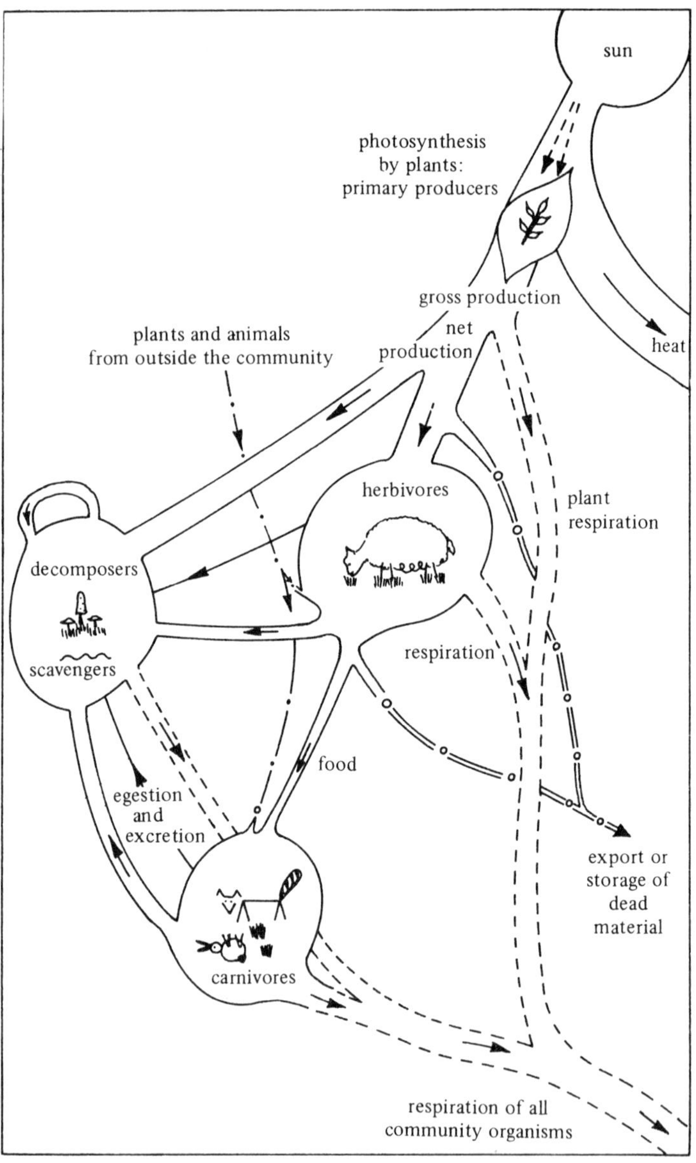

3.1 Energy entering communities

The sun is the only source of energy available to the world's living communities. Something of the order of 8.4 J cm^{-2} reach the outer atmosphere of the earth every minute. Very little of this **solar energy** (light energy) is available for transfer, through photosynthesis, to plant and animal tissue. A lot of it never reaches plants.

1 **Absorption** and **reflection** are the two ways in which solar energy is lost as it passes through the earth's atmosphere to the point where it can be used by plants. Consider the passage of light through the atmosphere to a plant. What things would absorb or reflect the energy on its way?

2 The actual amount of light energy reaching the outer atmosphere, and then the ground, depends on a number of factors which may vary during the course of a day or a longer period of time. Give any three of these factors.

Of the solar energy which is absorbed by plants, only as little as one or two to five per cent is used in photosynthesis. Even these figures are for times when the plant is photosynthesizing in favourable conditions. So much of the earth's surface has conditions which are far from **optimum** (the best possible) that the average amount of light assimilated for a whole year is nowhere near one or two per cent. Instead, it is only about one tenth of that.

3 What do you think happens to the light energy which falls on a plant and is absorbed but is not used in photosynthesis?

4 It is this very small amount of the available light energy, transferred to biochemical compounds in photosynthesis, which keeps all living organisms alive. Before you read on, make a list of the ways in which you think this energy can then pass through the living world. Fig 3.1 may help you. Remember that energy, unlike matter, cannot be recycled.

Fig 3.1 summarizes the movement of energy through and out of a community. It shows the **energy flow** through the community. In animals, the **consumers**, not all food ingested is assimilated and retained in the body tissues. It can be respired, egested or excreted, or used in reproduction. Once incorporated into animal tissue, it becomes available to other consumers or, on death, to **decomposers** and **scavengers**. But each time one organism becomes the food of another, a great deal of energy is lost. Nowhere near the amount which went into the making of the first organism is available to the animal(s) which consumes it, and even less actually becomes incorporated into the consumer. So only a minute proportion of the sun's energy eventually becomes fixed in a top carnivore, a relatively scarce organism in the ecosystem.

3.2 Primary productivity

Green plants are the **producers** in all ecosystems. By means of photosynthesis they make light energy available for themselves and for all **heterotrophic organisms**. Heterotrophic organisms or **heterotrophs** are those which cannot make their own food. Green plants are called **autotrophic organisms** or **autotrophs**. Basic or **primary productivity** is the rate at which plants assimilate solar energy. The word '**productivity**' means the same as the phrase '**rate of production**'. **Production** means what has been produced.

Primary productivity determines first the growth of vegetation in any one habitat and secondly the populations and productivity of the animals and other organisms within the community. Primary productivity is the base on which the whole community is founded. **Secondary productivities** are the rates of energy storage at consumer and decomposer trophic levels. (See also Problem 3.8.)

1 What external conditions limit primary productivity and the growth of plants? Give those which

a directly affect the rate of photosynthesis
b affect the rate of growth.

2 In cultivating the land we have increased the **yield** of our crops. Yield refers to the part of the crop plant which we harvest and use. It seems, however, from the data available that the maximum primary productivity (as opposed to yield) of crop plants has rarely been increased beyond that which can occur in man's absence (see Problem 4.1). What we have been able to do is to increase productivity in conditions where the maximum would not have been achieved and thus increase yield. In what ways may man affect the primary productivity?

3 Why do you think it is important to us to pinpoint **limiting factors** in productivity?

4 Which natural ecosystems do you think would have the highest primary productivity and why?

5 Both in natural ecosystems and in cultivated crops the consumer is concerned with the **net production** rather than with the gross production. **Gross production** is total assimilation during photosynthesis. What do you think net production is?

6 What sort of things could happen to the net production of a plant either during its lifetime or after?

7 The ratio of gross production to net production will vary during the life cycle of a plant. For example, in one young pine forest the ratio of gross to net production was calculated at 5:3. For a medium-aged pine forest it was 5:2.2. What would account for this relative fall-off in net production as the trees age?

8 One would also expect variations in the ratio of gross to net production in the life cycle of a herbaceous plant. Fig 3.2 shows respiration, biomass and gross production curves for a typical annual plant. Draw in the curve for net production.

Fig 3.2 Respiration, biomass and production curves for a typical annual plant

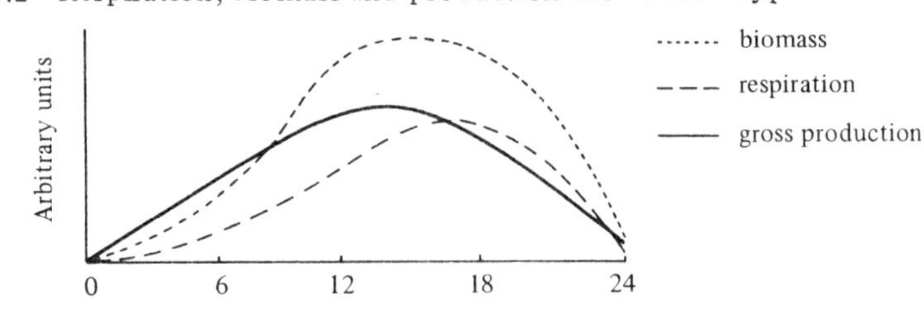

9 Of what use to us is it to know about the variations in net production?

3.3 Leaves and productivity

It is only the green parts of plants which photosynthesize. The primary productivity of a plant therefore depends on the amount of photosynthetic tissue (usually the leaves) and its structure and arrangement.

1 Fig 3.3 shows a plant growing with most of its leaves receiving as much light as possible. What kind of arrangements of leaves do you think would lead to maximum photosynthesis?

Fig 3.3 Leaf mosaic

Fig 3.4 Diagrams of plant shapes

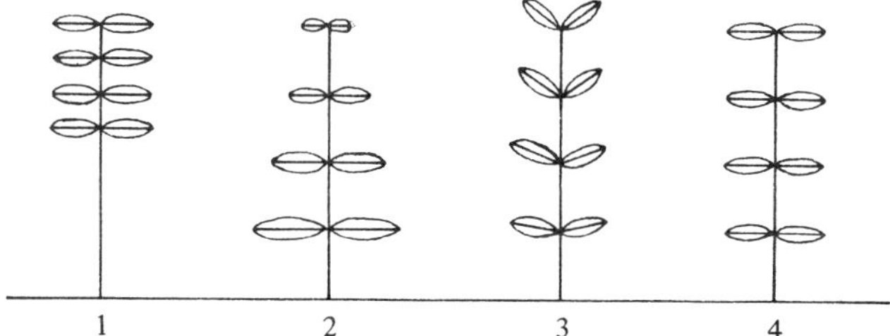

2 The diagrams in Fig 3.4 are of plants which all have the same leaf area index (see Problem 4.2) and the same leaf thickness, but different leaf patterns. Place these plants in order of potential productivity, with reasons.

3 How do you think the number of leaves will affect the productivity?

4 What effect would the length of life of the leaves have on productivity?

5 What kind of internal leaf structure will lead to high productivity?

3.4 Productivity round the world

Problem 3.2 touched briefly on the primary productivity of natural ecosystems. This problem, and Problem 3.5, look at some figures. What is primary productivity in terms of the mass of material produced, and by how much does it vary between ecosystems?

Fig 3.5 is a diagram of a cross-section of land and sea, showing the position of a number of different communities. What are considered to be reasonable estimates of gross productivity per day are given for three groups of communities.

Fig 3.5 Communities and primary productivity

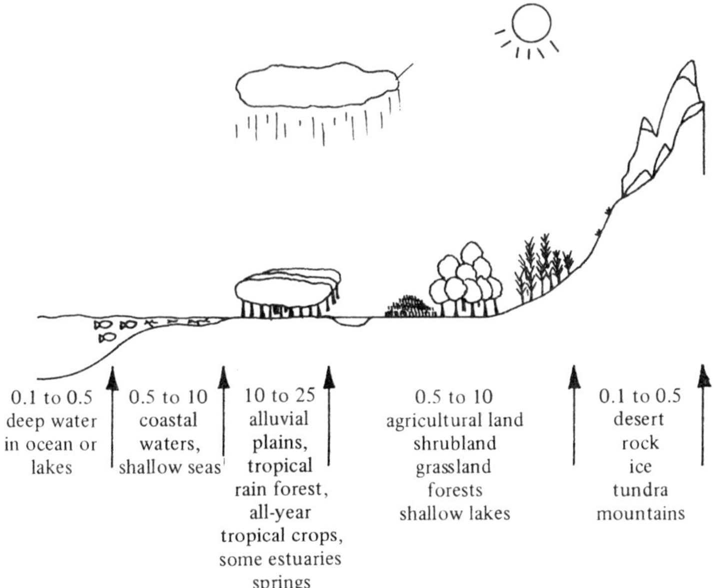

The figures are gross daily productivity/g m^{-2}

a Those where productivity is around 0.1 g to about 0.5 g or even up to 1 g for each square metre in one day: deserts, rock and ice, tundra, mountain vegetation, deep water in the oceans or in lakes.

b Those where productivity is from about 0.5 g or 1 g to 10 g per square metre: average agricultural land, shrubland, grassland, the sea around coasts, shallow inland waters.

c Those where productivity is from ten to twenty-five g per square metre: alluvial plains, tropical rain forests, year-round tropical crops such as sugar cane, coral reefs, tropical fresh-water reed swamps, estuaries, mineral springs.

1 What external conditions contribute to the high productivity of the ecosystems in group **c**?

2 What are the limiting factors to production in

a deep water, **b** average agricultural land, **c** tundra?

3 What types of plant contribute to the primary productivity in the ecosystems mentioned?

3.5 Productivity and ecosystems

When considering differences between ecosystems, one should ideally compare figures of net productivity and not of gross productivity. There is, for example, a much bigger difference between the gross and net annual productivity figures for a tropical forest than between gross and net annual productivity for a temperate deciduous forest.

1 Why do you think this is so?

Table 3.1 gives estimates of the annual net primary productivity for a number of different ecosystems.

Table 3.1 Annual net primary productivity estimates for various habitats. Normal ranges of values vary depending on the category, but the productivity of most habitats is within one-half to twice the mean value given.

Habitat	Annual net productivity estimates/g m^{-2}
Terrestrial	
Extreme desert, rock and ice	3
Desert scrub	100
Tundra and alpine	140
Temperate grassland	500
Savanna, shrubland	600
Boreal forest (coniferous)	800
Temperate deciduous forest	1200
Temperate coniferous forest	2800
Tropical forest	5000
Agricultural land	
Annual crops (temperate)	2200
Annual crops (tropical)	3000
Perennial crops (temperate)	3000
Perennial crops (tropical)	7500
Swamps and marshes (emergent vegetation)	
Salt marsh	3000
Fresh-water reedswamp (temperate)	4300
Fresh-water reedswamp (tropical)	7500
Fresh-water (submerged vegetation)	
Attached plants (temperate)	600
Attached plants (tropical)	1700
Marine	
Attached algae and estuaries	3200

2 a Which of the communities in the table are the most productive? What external conditions contribute to their high productivity?

b Which communities are the least productive, and why?

3 Marsh plants are highly productive. What features of their habitat and habit encourage a high productivity?

4 The figures given in the table may be obtained by collecting samples of the above-ground parts of the plants at the end of the growing season, and then finding their dry mass. How does this method give an incomplete picture of production?

5 Apart from determining dry mass, in what other ways do you think productivity may be measured?

3.6 Nutrients and thermoclines: productivity in the sea

Table 3.2 gives estimated annual net productivity in some marine ecosystems.

Table 3.2 Estimated net productivity in some marine ecosystems

Ecosystem	Estimated annual productivity/g m^{-2}
Temperate seas	170
Tropical seas	40–120
Upwelling area of a tropical sea	200–700
The sea around coasts, continental shelves	200–1000
Estuaries	500–6000
Coral reefs	3000–5000

1 Of the marine ecosystems listed, some estuaries have the highest productivity. What factors do you think might account for this?

In the open sea the primary producers are the phytoplankton which occupy the surface layers of the sea where light penetrates. In temperate seas, as the water warms up every year, a **thermocline** forms. It gives an upper layer of warmer water (called the epilimnion) lying on water (the thermocline) which shows a rapid fall in temperature. Underneath the thermocline lies the great mass of cold water (the hypolimnion).

Each year when this situation has established itself there is hardly any circulation of water across the thermocline. As individual organisms in the plankton die, their remains sink through the thermocline, and there is a rapid loss of nutrients from the upper layer. In the autumn, as the epilimnion cools, the thermocline breaks up, often helped by storms. The level of nutrients rises again in the surface layer as nutrients come from the hypolimnion and circulate once more.

Fig 3.6 Phytoplankton and nutrients

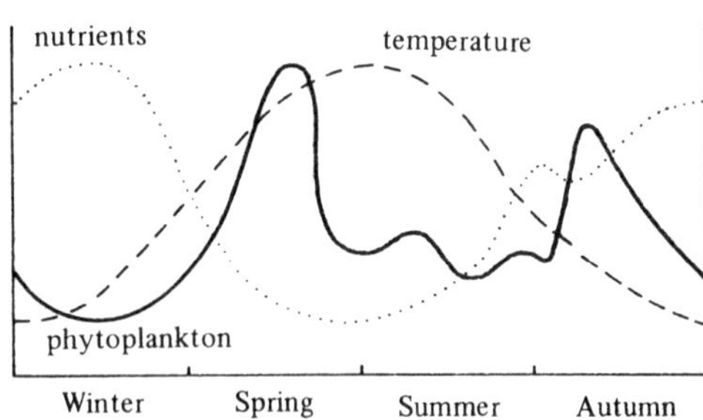

Fig 3.6 shows typical changes in phytoplankton numbers and nutrients during the year in temperate seas.

2 What factors may account for

a the spring increase in phytoplankton
b the rapid decrease of phytoplankton in the summer
c the increase of phytoplankton in the autumn?

3 In tropical seas where light and temperature are favourable all the year, primary productivity is lower than in temperate seas. Why do you think this is? Compare the figure in Table 3.2 for the upwelling area of a tropical sea where currents and winds cause cold water to move up to the surface with that for a tropical sea with no upwelling.

4 In arctic seas productivity is also lower than in temperate seas. Why do you think this is?

3.7 Ecological efficiency

Fig 3.7 shows the energy flow for the community in a hypothetical, small, but rich pond where there are many and various organisms. Most of the food is produced inside the ecosystem, but a little comes from outside. The initial 4% gross production is high.

1 List the ways in which energy is lost from the system.

2 The **ecological efficiency** is the efficiency with which production at each trophic level is converted into production at the next higher level. What is the ecological efficiency of the ecosystem shown here? In other words, what is the net production at each level as a percentage of that taken in?

(Compare your answer with a farm animal selected for high yield in relation to input. Pigs are thought to be the most efficient converters of food to edible products. Up to 20% of the energy fed to a pig can be used by us as meat or other products. The other 80% goes in pig respiration, egested and excreted matter, and inedible parts of the body such as the tail.)

3 What percentage of the net plant production goes through the decomposer chain rather than through the grazing chain?

4 Consider a pond, or some other similar closed ecosystem. What do you think is meant by 'import into' and 'export out of' the ecosystem?

3.8 Consumers and food chains

Consumers and **decomposers** in an ecosystem are **heterotrophic organisms**: they cannot make their own food. They have to obtain their food, in one way or another, from green plants. Most heterotrophic organisms fit easily into the category of consumer or decomposer.

1 How would you describe each of these two categories, and what type of organism would you put into each?

Within the consumer group, one can distinguish a number of feeding methods relating to different types of food.

2 How would you describe each of the groups in the following list of heterotrophic organisms? Herbivore, carnivore, omnivore, detritivore, scavenger, parasite.

The animals in each group occupy different places in a community. They may form different **trophic levels**, which indicate their place in the energy flow through a community. Organisms whose food is obtained from plants by the same number of steps belong to the same trophic level. Thus **herbivores**, feeding directly on the primary producers, are **first level consumers** (or **primary consumers**). **Carnivores**

Fig 3.7 Energy flow through a hypothetical pond

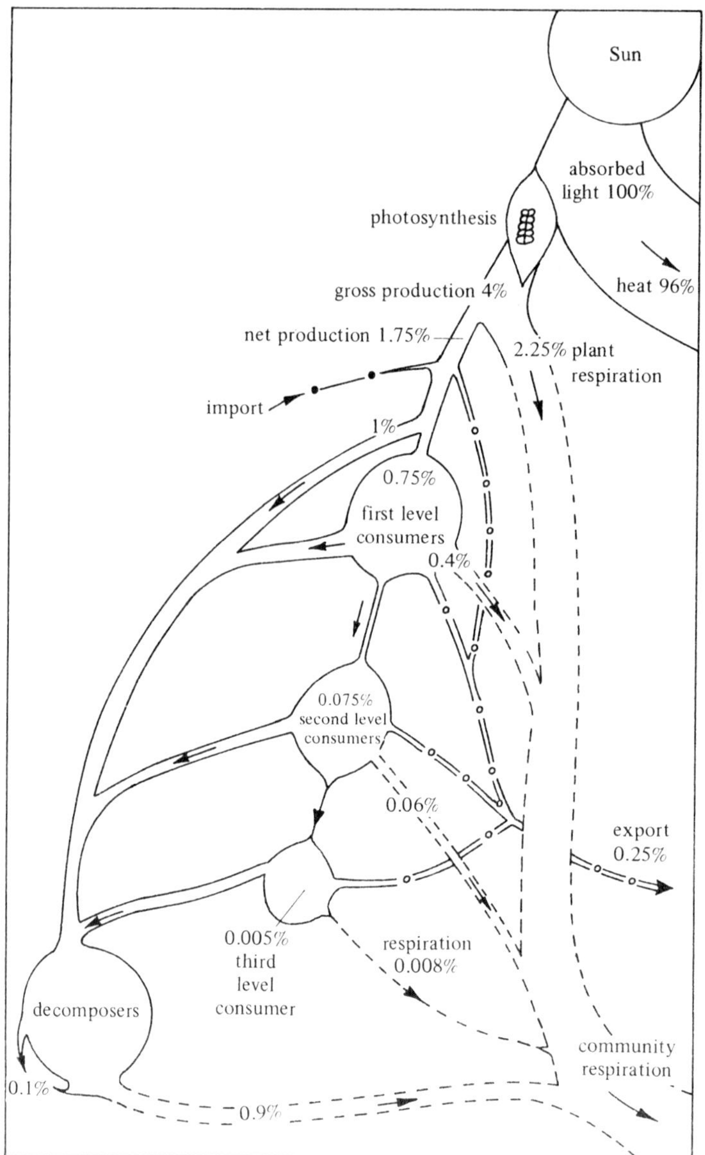

which feed on the herbivores are **second level consumers** (or **secondary consumers**). The next, and usually last, level is **third level (tertiary) consumers**.

The word **niche** is sometimes used to describe an organism's position within the community, and the part it plays there. Organisms occupy different niches depending on their structure, physiology and behaviour. An organism's ecological niche depends not only on where it lives (its habitat), but also what it does there. The American ecologist E. P. Odum describes the habitat as the organism's address, and the niche as its profession.

Fig 3.8 Levels of organisms

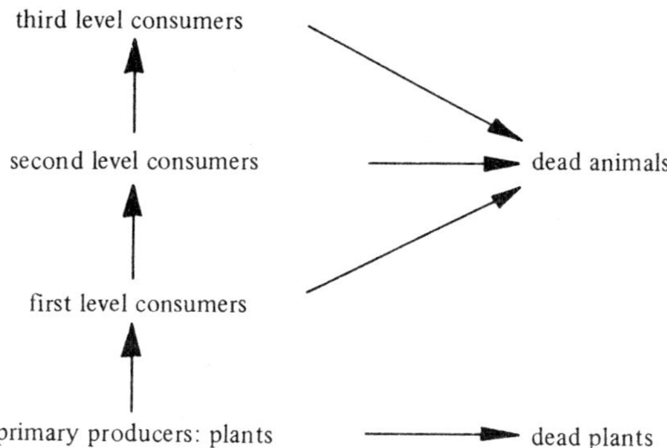

3 Place the groups of organisms from question 2 at their correct level on Fig 3.8.

4 Now insert the names of different plants and animals in place of the producers and consumers on the diagram. This will give you a **food chain**. A very simple food chain is grass, cow, man.

In fact, there are not many animals which feed on only one, or a few, organisms. It is hardly possible to analyse an ecosystem in terms of food chains and straightforward trophic levels. Most animals eat a variety of food. A carnivore may be both a second and third level consumer and possibly also a detritivore or may occasionally eat plant material; tadpoles eat different food from frogs, caterpillars from butterflies, and so on. Instead of food chains it is nearer the true situation to represent feeding relationships within a community as a **food web**.

5 Construct a food web for any community familiar to you, or involving some of the food you have eaten today.

6 The energy which is available for consumers and decomposers comes from plants. What consumers ingest represents the amount of energy which enters the first consumer level, some of which becomes available for higher levels. Only a small part (around ten per cent) of the food eaten by a first level consumer does become available for higher levels. What happens to the rest?

3.9 Decomposers

Much of the energy entering each trophic level leaves it through the **decomposer** pathway, a less conspicuous pathway than the more widely understood one of herbivore and carnivore. The dead remains of animals and plants, as well as excreted and egested matter, can all be metabolised to provide energy. Some carrion eaters and scavengers such as blowfly maggots, crabs, vultures, hyenas or crows are large or conspicuous. The name given to biological material in various stages of decomposition is **litter** or **detritus**. The term litter usually refers to material in terrestrial ecosystems, and detritus is more generally used when referring to aquatic ecosystems.

Litter and detritus are broken down and used as food by earthworms, small arthropods, other invertebrates, and especially by micro-organisms, in particular bacteria and fungi. The less digestible the material produced by a community, and wood is an example, the more important is the decomposer pathway in the food web.

1 Why do you think the decomposer pathway is important?

The processes involved in decomposition are physical and biological. The material being decomposed may be leached by the rain, and then fragmented by the action of wind, heat and **detritivores** which feed on dead remains and organic debris. At the same time, or soon after, fungi and bacteria, with their extra-cellular digestion, decompose the material chemically. The process goes on until either the material has completely disappeared, or only the very resistant parts such as bones are left.

Most of the information on litter feeders comes from work on the breakdown of leaf litter in woodlands and forests. In temperate deciduous woods, trees lose their leaves in the autumn. In tropical evergreen forests, leaves drop to a certain extent throughout the year. The leaves and other parts of the plant fall to the ground and form a layer of litter. It is estimated that probably less than 10% of the living leaves in a forest are normally eaten by herbivores, ranging from caterpillars to birds or mammals. This gives about 90% of leaf production which could enter the decomposer pathway.

One study, on the breakdown and decomposition of leaf litter, done in a Hertfordshire wood, used discs of 2.5 cm diameter cut from oak and beech leaves. 50 leaf discs per bag were put into nylon mesh bags, using mesh of different sizes for each set of bags. The bags were buried 2.5 cm deep in newly dug soil, in July. Every two months the leaf discs were measured to see how much of the leaf area had disappeared. The mesh sizes used were:

7.0 mm which allowed all micro-organisms and invertebrates in the soil to enter the bag
1.0 mm which allowed all micro-organisms and invertebrates except earthworms to enter
0.5 mm which allowed only micro-organisms and small invertebrates to enter
0.003 mm which allowed only micro-organisms to enter.

Table 3.3 gives the results of the investigation for the oak leaves.

Table 3.3 Disappearance of leaf discs/%

	7 mm mesh	0.5 mm mesh	0.003 mm mesh
July	0	0	0
August	20	6	0
October	70	19	0
December	86	33	0
February	91	37	0
April	93	39	0

2 Which are the most important animals in the breaking down of leaves?

3 a Is there any seasonal variation in the rate of breakdown?
b List any three environmental factors which will affect the rate of breakdown. Say why and how they do.

4 In the 0.003 mm mesh bag, the leaf area of the discs was the same after nine months as at the start, but the dry mass had decreased. Say why you think this was so.

5 What other changes would you expect there to be in the leaf discs after they had been buried?

6 In a woodland such as the one in which this investigation was done, each year's litter disappears within a year. In some communities, however, litter accumulates. In what communities would you expect to find a large amount of litter?

3.10 Pyramids

Within a community the food relationships and trophic levels can be shown by one or more **ecological pyramids**, as well as by food webs. The producer organisms (plants) form the base of the pyramid, and the rest is made up of successive consumer levels.

Three main types of pyramid are described below. Theoretical diagrams, much simplified, are given in Fig 3.9.

A **pyramid of numbers** shows the number of organisms at each trophic level at any one time. The producers at the base of the pyramid are abundant, the first level consumers are relatively abundant, second level consumers less so, and any third level consumers at the tip of the pyramid are relatively few. In this type of pyramid any one individual counts for the same as any other. A tree and a grass plant, a cow and a grasshopper would each count as one.

A **pyramid of biomass** is based on the total amount of living material at each trophic level at any one time. It shows the **standing crop** or standing biomass present at a given time. Do not confuse it with productivity. A fertile, productive field which is being constantly grazed by cows will have a high productivity, but a low standing crop. A field too poor to be grazed has low productivity but possibly a higher standing crop at a particular time because of the uncropped grass.

A **pyramid of energy** shows the rate of energy flow, or productivity, at successive trophic levels. Usually productivity cannot be assessed by counting or weighing as in the other two. One needs also to know the rate of production and consumption of numbers or mass. Numbers and biomass of consumers depend not on the numbers and biomass of producers and lower level consumers at any one time, but on the **rate** at which food is produced. Producers might be few or small, but at the same time they could be being produced at a massive rate. A pyramid of energy therefore considers the time factor. It shows the amount of new tissue or organisms (as grammes, or as joules which is the energy stored) produced in a unit of time.

Decomposer levels are not often shown on pyramids.

Fig 3.9 Pyramids

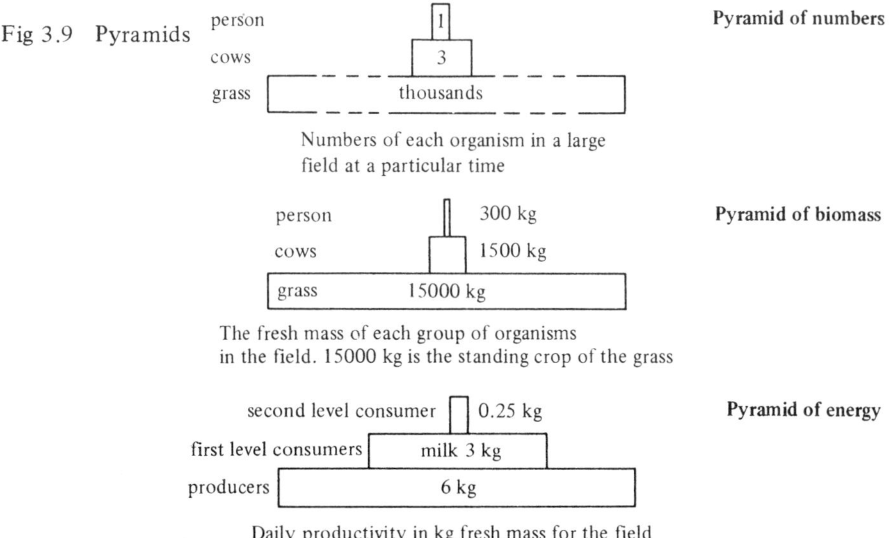

1 As suggested above, for some communities the pyramid of numbers may be inverted. The base may be smaller than one or more of the upper levels. Give an example of such a pyramid.

2 Can the energy pyramid ever be inverted? Explain your answer.

3 Which of the three pyramids do you think is the most useful one in giving the best picture of the community, and why?

4 Fig 3.10 gives some pyramids of numbers. Match each pyramid in Fig 3.10 to one of the following:

a grass, zebra, lion
b rosetree, aphids, parasites of aphids
c tree, insects in the tree, birds
d phytoplankton (which may be few in number but with a high reproductive rate), zooplankton, fish
e the organisms living in a cave, or in the depths of a lake.

Fig 3.10 Pyramids of numbers

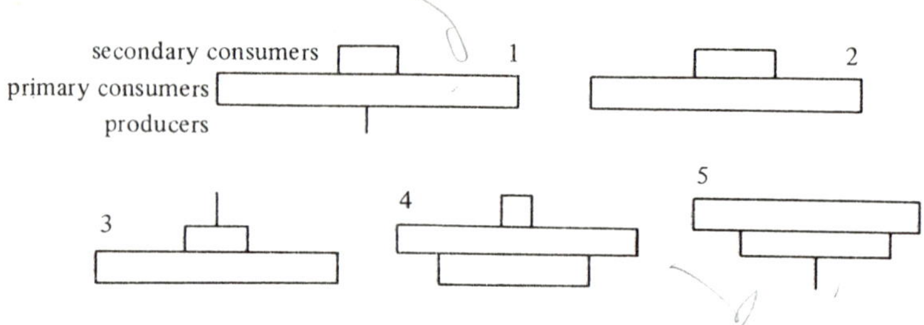

3.11 Man as a consumer

1 Bearing in mind the very great loss in energy as one proceeds up the food chain, which would be the most economical for us to use as food: plants, herbivores or carnivores?

2 Are the consumers in your diet mainly first, second or higher level consumers?

3 What objections are there to using only plants as food?

4 How do you think we could make more use of the sunlight which reaches the earth's surface?

5 Which of the following is the more efficient converter of one tonne of animal food to human food?

1 bullock eating the tonne in 120 days, producing 100 kilogrammes of beef.
300 rabbits eating the tonne in 30 days, producing 100 kilogrammes of rabbit flesh.

6 An animal grows faster when young than when adult. In chickens, for example, growth slows down between three and four months after hatching. Hens do not lay eggs until they are between five and six months old, when growth has almost stopped. What is the best time to kill a food animal from the point of view of efficiency of conversion of producers to human food?

7 How can we help to avoid the loss of energy which occurs between plant production and this production reaching our herbivorous domestic animals or us as primary consumers?

8 Discuss possible new sources of food for people.

4

CROP PRODUCTION

4.1 Crop productivity

In experimental crops, where the plants are growing quickly in ideal conditions, they may lose in respiration as little as 10% of their gross production. Normally, however, plants lose more like 30% to 50%. Following from the first figure, it is sometimes stated that 90% of what crop plants produce could be harvested. To do this, individual plants just past their peak rate of photosynthesis would have to be replaced by others approaching their peak rate. This is possible with algae, but even then it increases productivity only to a limited extent because of the high respiratory rate of small organisms.

Table 4.1 shows the variation in net productivity for different crops, and for the same crop as a world average and in an area of high productivity.

Table 4.1 Annual and daily net primary productivity of some crops/g m^{-2}

	World average			Average in area of highest yield		
	Per year	Per day	Per day of growing season	Per year	Per day	Per day of growing season
Wheat	344	0.94	2.3	1250	3.4	8.3
Oats	359	0.98	2.4	926	2.5	6.2
Maize	412	1.13	2.3	790	2.2	4.4
Rice	497	1.36	2.7	1440	4.0	8.0
Hay	420	1.15	2.3	940	2.6	5.2
Potatoes	385	1.10	2.6	845	2.3	5.6
Sugar beet	765	2.10	4.3	1470	4.1	8.2
Sugar cane	1725	4.73	4.7	3430	9.4	9.4
Intensive algal culture				4530	12.4	12.4

1 How do you account for the difference between the values for the daily productivity when taken as an average over the whole year (columns 2 and 5) and during the growing season (columns 3 and 6)?

Large areas considered together have a lower rate of production than that from a small, well-populated country with a good climate and intensive cultivation in fertile soil only. In Table 4.1, the average world productivity for all crops is considerably lower than the averages from areas with a high rate of production. The figures in the last three columns come from the Netherlands (wheat, potatoes and sugar beet), Denmark (oats), Canada (maize), Italy and Japan (rice), California (hay), Hawaii (sugar cane), Tokyo (algae).

2 Discuss the measures which could be taken to eliminate the differences apparent

from the table.

When one considers figures of primary productivity of cultivated ecosystems and the more productive natural ecosystems, we seem not to have increased maximum primary productivity beyond that which may occur in natural ecosystems. Of course, if water and nutrients are limiting crop growth, then irrigating and fertilizing the land will increase it. But look at Fig 3.5 in Problem 3.4 which gives figures for world-wide gross productivity. Divide these figures by two to give a very approximate figure for net productivity.

3 How do the world-wide figures compare with those in the above table?

4.2 Leaves and yield

If the plants in a crop are widely spaced each individual plant may achieve its maximum yield, but the yield of the whole crop may be lower than if plants are grown closer together.

1 Why do you think this is?

2 On the other hand, if plants are very closely spaced, total crop yield may be high but none-the-less lower than could theoretically be possible at that spacing because of the activities of other organisms. Explain why this should be so.

The **leaf area index (LAI)** of a plant or crop is the ratio of its total leaf area to ground area covered. The LAI gives an indication of the efficiency of light interception. For example, a plant such as maize which has leaves at right angles, or greater, to the stem (the leaves are horizontal or drooping) has a total leaf area of about four times the actual ground area covered by the leaves.

3 What is its approximate LAI?

4 With increasing density of a crop, the LAI is increased. If the LAI were, say, doubled so that the ground area was covered eight times by leaves, would you expect the yield to be increased?

Plants with more upright leaves than maize can be planted densely without decreasing the yield per plant, thus giving a higher total crop yield. Indeed, some of the newer rice and wheat varieties which have a greatly increased yield have been bred to achieve this kind of shape.

5 Say why you think such a shape gives a higher yield of grain.

On average, crops grown in developed countries yield about four times more than the subsistence agriculture of less developed countries. It is interesting to consider that part of the difference is due to a kind of 'energy subsidy' given to the higher yielding crops.

6 What form does this energy subsidy take?

4.3 Leaves and the growing season

In spite of man's help, cultivated ecosystems may not have a higher daily productivity than some natural ecosystems (see Problem 4.1). Increasing the **photosynthetic efficiency** of crop plants, if that is possible, would be one way to increase productivity (see Problem 4.2 for some discussion of this). Of equal importance, and perhaps easier to achieve, is the possibility of extending the **period of maximum growth**. Many crops have a high rate of production for only part of the year.

Extending the period of maximum growth would make more use of the available light energy.

Look first at Fig 4.1 which shows the general trend of yield of irrigated and fertilized agricultural crops related to the total photosynthetic surface per unit area of land, that is, the **leaf area index**, LAI (see Problem 4.2).

Fig 4.1 Leaf area index and productivity

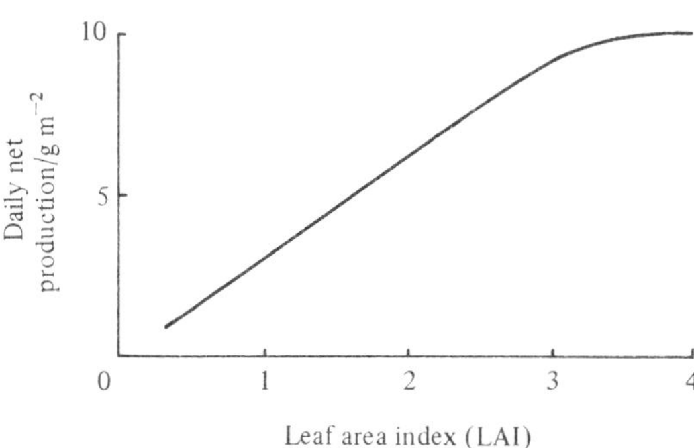

1 What conclusion can you draw from this figure?

Examples of LAI which give maximum rate of dry matter production are between 3.5 and 4 for kale, about 6 to 7 for sugar beet, 2 for cotton. Beech may have a LAI up to 8. This gives a dense shade in beech woods and there is very little ground flora because of the low light intensity under the trees. In general, a LAI of between 3 and 4, reached as quickly as possible after germination and remaining at that value for most of the growing season, gives the highest yields. A LAI of 5 or 6 to 10 includes leaves which add nothing to the plant's production and are a drain on its resources.

The efficiency of the photosynthetic system can be expressed as the mean rate of dry matter production per unit leaf area. This is called the **unit leaf rate (ULR).** In a crop ULR averages about 4 g to 6 g for each square metre per day, but can vary between 2 g and 14 g depending on the conditions. If a crop can achieve a ULR of 5 g per square metre per day at the beginning of the growing period, rising to about 10 g when external conditions are at their best, very high yields are possible (see column 6 in Table 4.1, Problem 4.1).

Consider any common agricultural field crop in England. Exclude grassland, orchard crops or forests. The seeds are sown in spring, hopefully to miss the worst frosts, about three months after the darkest time of the year. They germinate and start to photosynthesize as the days are rapidly lengthening. Their leaf area increases slowly at first and then more rapidly. By early May they have achieved half the maximum midsummer ULR. By mid-June the plants are growing quickly. There is usually enough rain, the days are long and warm and the light intensity is high. Photosynthesis proceeds apace and the plants grow rapidly, probably achieving the optimum LAI and the maximum ULR.

By July, however, in many cases well before the crop is ready, the days are becoming shorter and photosynthesis takes place for slightly less time each day. By the beginning of September the ULR falls to half maximum. In some crops (cereals, potatoes), the increase in LAI ceases at flowering, no new leaves grow and old ones die. Spring barley is a crop which has one of the shortest times with a full green canopy. Crops such as sugar beet and kale, however, have a longer period of high LAI as new leaf production balances the death of old leaves. In some crops LAI may therefore not reach its maximum until after midsummer when external factors become steadily less favourable for photosynthesis.

The period of maximum productivity, then, is fairly short. As the days are lengthening in late spring and early summer, the LAI is low because low temperatures result in a relatively late germination. As the days get shorter, LAI is at its optimum. The crop may then be harvested in August or September, while there are still many weeks of longish days but no leaves to use them.

2 Discuss how the period of maximum productivity for crops could be lengthened. Some points are raised in the following paragraphs.

Light intensity usually reaches maximum for full photosynthesis early on a summer day, indicating that photosynthesis could proceed at maximum rate at a much lower light intensity than is finally obtained by mid-day. Indeed some plants (e.g. sugar beet) orientate their leaves at an angle to the sun during the period of maximum light intensity, and more light passes to the lower leaves. If plant breeders could produce varieties of other crops which do this, a very much higher ULR could be obtained.

Both changing the colour of the soil to increase the proportion of radiation absorbed (a thin dressing of soot can increase soil temperature), and sloping the soil surface towards the mid-day sun, could raise soil temperature in the crucial early days of germination and leaf expansion, and thus bring forward the date at which seedlings can be established successfully. Experiments have shown that ridging the soil to give south-facing and west-facing slopes on which seed is sown can greatly speed up seedling establishment. As an example, in one experiment, the time from sowing to emergence of maize was reduced by about twenty-five days.

4.4 Factors affecting leaf area index and unit leaf rate

The amount of dry matter produced by a crop depends on the size and efficiency of the photosynthetic system. The leaf area index (total area of leaves per unit area of land, Problem 4.2) gives the size of the system. The unit leaf rate (mean rate of production of dry matter per unit leaf area, Problem 4.3) gives the efficiency. The highest rate of dry matter production is when the product of LAI and ULR is greatest: crop growth rate = LAI × ULR. After an optimum, as leaf area index rises, the unit leaf rate falls because shading reduces the average amount of light per leaf.

Any factor which changes the amount of dry matter produced will do so by changing either one or both of the leaf area index and the unit leaf rate.

1 Below is a list of some of the factors which affect the rate at which plants increase in mass. For each, say whether you think it affects the leaf area index or the unit leaf rate (or both), and how.

fertilizer	light intensity	temperature
water	light duration	carbon dioxide

improvement in crop husbandry (includes soil preparation, weeding, crop rotation)

2 Consider your answer to **1**. Do you think that an increased yield of crop plants is likely to come from an increase in the size or the efficiency of the photosynthetic system?

3 Distinguish between the **economic yield** and the **biological yield** of a crop plant.

4 If the biological yield of the whole plant can be increased, to which branch of biological science must we turn in order to channel the biological yield into an economic yield?

4.5 Factors affecting plant productivity

A tall leafy crop plant is investigated to determine its net primary productivity. It is grown under ideal conditions for maximum photosynthesis.

Which of the following factors would affect the net primary productivity of this plant over a period of a day?

- **a** The position of the leaves relative to the stem: for example, bunched at the top or spaced down the stem
- **b** The angle of the leaves to the stem
- **c** The rate of respiration
- **d** The leaf area index
- **e** The rate of transpiration
- **f** The unit leaf rate
- **g** The arrangement of the leaves round the stem
- **h** The amount of water absorbed
- **i** The amount of photosynthetic product stored
- **j** The amount of fertilizer in the soil
- **k** The presence of photosynthetic tissue in addition to leaves
- **l** Leaf structure
- **m** The age of the leaves
- **n** The length of the light period.

4.6 What makes a barley grain?

In cereals, the major economic part of the plant is the grain. The straw, which is the dead leaves and stems, is also useful. The grains are produced on an ear (Fig 4.2). The larger grains are at the bottom and grain size decreases upwards to the top of the ear. The grains grow at different rates and may differ considerably in their final mass. For a long time it was thought that the photosynthesis of all the green parts of a plant during the season of grain production would contribute to the final economic yield. Evidence now points to other mechanisms. In barley, for example, a lot of the starch in the grains is formed by photosynthesis in the ear itself, not in the leaves.

Fig 4.2 Barley ear

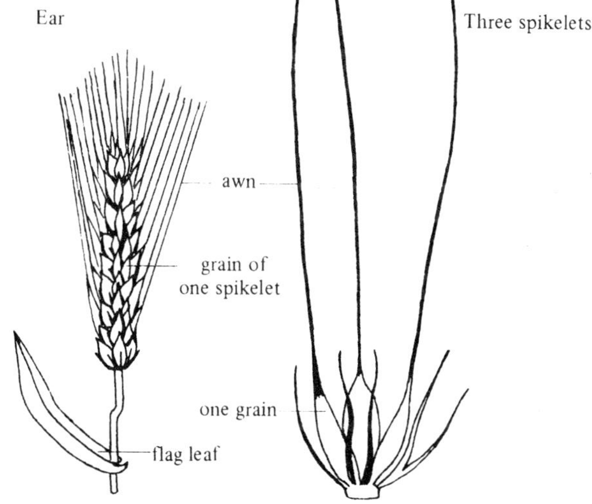

If you look at a barley (or wheat) flowering stalk about the time the ear starts to grow, you will see that the leaves are past their prime. The amount and duration of photosynthesis in the ear and its flag leaf is therefore probably very important in determining the final yield of grain in our two main cereal crops.

1 How would you attempt to find out whether it is the leaves or the green tissue of the ear which contributes most to the growth of the grain?

In fact, the main source of carbohydrates for the developing grains seems to be the photosynthetic tissue of the ear itself and the flag leaf. The flag leaf is just below the ear and is the last leaf to grow on the flowering stalk. In one investigation $^{14}CO_2$ was supplied to the flag leaf. When the resulting radioactivity of the grains was measured, the results shown in Fig 4.3 were obtained. Fig 4.4 (lower curve) shows the growth rate of individual grains. On the horizontal axes of both graphs the low numbers represent grains at the base of the ear and the higher numbers those at the tip.

Fig 4.3 (left) The distribution of radioactive photosynthetic products between barley grains
Fig 4.4 (right) Awn lengths and growth rate of barley grains

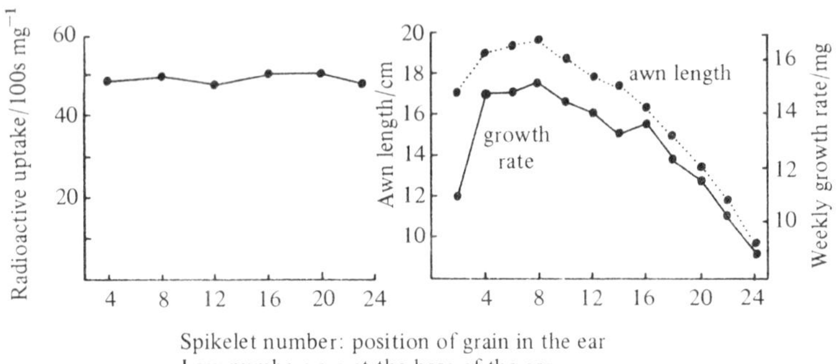

Spikelet number: position of grain in the ear
Low numbers are at the base of the ear

2 When you compare Fig 4.3 with the lower curve on Fig 4.4, what conclusion can you draw about the contribution of the flag leaf to grain growth? Do you think the flag leaf has contributed unequally to the mass of different grains?

In an attempt to see what other tissue was contributing to the growth of the grain, the investigators suggested that the awn at the tip of each grain contributed to its own grain, and only to that one. They also suggested that the longer the awn the bigger the grain eventually became. The upper curve on Fig 4.4 gives the awn lengths of different grains up the ear.

3 What evidence is there from Fig 4.4 to support the awn hypothesis?

In an investigation to test the awn hypothesis, single awns were supplied with $^{14}CO_2$. After three and a half hours the amount of radioactivity in each grain was measured. For the grain at spikelet 10, twenty-one days after pollination, of the ^{14}C supplied 99.6% was in the grain, 0.3% had moved up the ear, and 0.1% had moved down the ear.

4 How do these results also support the awn hypothesis?

You may meet the words **sources** and **sinks** in your reading. The producing organs are the sources. The places where the products of photosynthesis are stored or used are the sinks.

5 In this problem, which is or are the source(s) and which the sink?

6 When the sink is of economic importance, a knowledge of the source is useful. Once we have this knowledge, what steps could be taken to use it?

4.7 Photosynthesis during one day

The rate of photosynthesis of a crop plant was investigated during the course of a warm, sunny English summer day. The method used was to determine the dry mass of leaf discs cut from the leaves at hourly intervals. The results are shown in Table 4.2.

Table 4.2 Dry mass of leaf discs

Time of sampling	Mass of discs/g	Time of sampling	Mass of discs/g
4.00 hours	3.013	14.00 hours	3.728
5.00	3.018	15.00	3.795
6.00	3.026	16.00	3.870
7.00	3.040	17.00	3.948
8.00	3.076	18.00	4.001
9.00	3.161	19.00	4.021
10.00	3.271	20.00	4.021
11.00	3.402	21.00	4.004
12.00	3.541	22.00	3.958
13.00	3.662	23.00	3.860

1 Some details of experimental design are omitted from the introduction given above. Give more details of the experimental method which you think was used.

2 Present the data graphically to show most clearly the changes during the day.

3 Analyse your graph, and account fully for the changes during the course of the day.

4.8 Weeds

It has been estimated that approximately 10% of the possible production of crops is lost because of weeds. In natural communities there are no weeds. A plant is only a weed when it is growing in a place we do not want it to grow.

1 In what ways do you think weeds can affect the growing and the harvesting of crops?

2 What farming, or gardening, methods do you think encourage the germination and growth of weeds?

3 What farming, or gardening, methods could be used to help prevent weeds?

4 Tables 4.3, 4.4 and 4.5 give data from experimental treatments of two crop plants. They show the effect of weeds on the final yield of the two crops, red beet and carrot. Consider these tables in the light of your answer to **1**. What conclusions may be drawn from the tables?

Table 4.3 The effect of the position of weeds on the yield of red beet/kg m^{-1} row

	Weed free	Weeds left in crop row	Weeds left between crop rows
Marketable beet	1.5	1.2	0.9
Weeds	0	0.2	0.3

Table 4.4 The effect of time of weed establishment on mass of carrot root/g. 37 carrot plants and 16 weed plants in each container.

Treatment	Mean mass per root/g
No weed sown	156.6
Carrots sown two weeks before weeds	126.3
Carrots sown one week before weeds	106.3
Carrots and weeds sown on same date	60.7

Table 4.5 The effect of density and weeds on yield of carrot/kg m^{-2}

	Rows separated by	
	60 cm	30 cm
Marketable carrots	2.4	3.4
Weeds	0.8	0.4
Cover carrots/%	63.4	90.9
Cover weeds/%	14.0	5.1

5 Consider the yield of red beet with and without weeds in Table 4.3. Discuss whether it would be worthwhile financially to weed the fields where the red beet is growing. You will have to estimate the cost of labour, machines, chemicals, the price of red beet, the size of the fields and so on. There is no need to aim at an accurate balance sheet. Simply try to get some idea of all the financial factors involved before even a relatively easy decision such as this can be taken.

4.9 Potato blight

"A fatal malady has broken out amongst the potato crop. On all sides we hear of the destruction. In Belgium the fields are said to have been completely desolated. There is hardly a sound sample in Covent Garden Market. The disease consists in a gradual decay of the leaves and stem, which become a putrid mass, and the tubers are affected by degrees in a similar way. The first obvious sign is the appearance on the edge of the leaf of a black spot which gradually spreads; the gangrene then attacks the haulms, and in a few days the latter are decayed, emitting a peculiar and rather offensive odour. When the attack is severe the tubers also decay."

So wrote the Editor of *The Gardener's Chronicle and Agricultural Gazette* in August 1845, a year which saw the start of the Irish famine caused by the loss of the potato crop due to potato blight. One million people died from starvation or diseases resulting from near-starvation, and a further one and a half million emigrated.

Potato blight is caused by a fungus, *Phytophthora infestans*, which attacks potato plants, causing the leaves, stems and tubers to be destroyed. The life cycle is given in Fig. 4.5. If an infested potato tuber is planted in the spring, the fungus grows up from the tuber into the shoots growing from it. The fungus kills the infected cells of the potato plant and causes spots (called lesions) on the leaves. Sporangia are produced which can themselves infect fresh potato tissue, or they can produce motile zoospores which also infect new places.

Fig 4.5 Life cycle of *Phytophthora infestans* (sexual stage omitted)

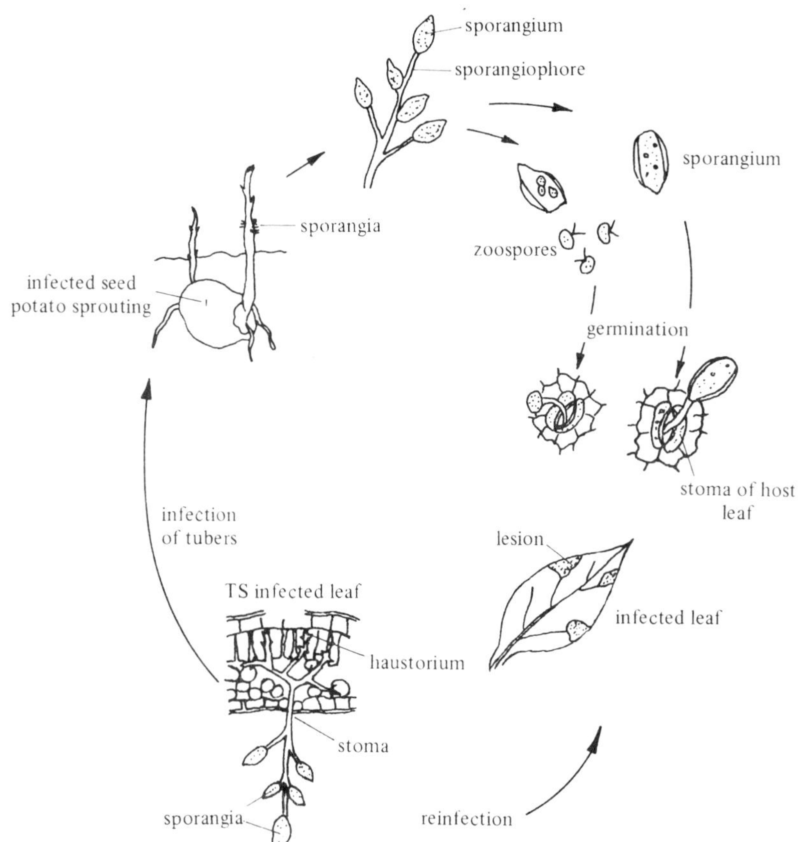

The disease is carried on from one year to the next by infected tubers. They may be throwouts, and dumps of unwanted potatoes can be a source of blight. Or they may be among those left in the ground from the previous year's crop, infected in the soil by spores washed down from blighted leaves or stems. Once established in the shoots, even from only a few infected tubers, the disease can spread in the right conditions.

Infected tubers have small, sunken watery spots at first, which turn brown and may be secondarily infected by other fungi and bacteria. They are inedible.

The main potato crop in England and Wales is planted in April and the potatoes lifted in September or October. In the warmer parts of the country, and the Channel

Islands, early varieties of potatoes are planted, yielding new potatoes from May. Blight occurs mainly in the summer. A very high relative humidity is necessary for the production of many sporangia, which die within two hours if the air is dry. Spores and sporangia will germinate on leaves only if a film of water is present. Zoospores need two and a half hours of cool (10 to 15 °C), wet weather to germinate. At slightly higher temperatures (up to 21 °C) the fungus will grow and form sporangia very rapidly. New lesions occur within three days at 20 °C, within six days at 15 °C and within thirteen days at 11.5 °C.

1 What kind of weather do you think would favour a blight epidemic?

If the leaves and stems of the plant are killed by the end of July, up to half the potato crop may be lost. Much more commonly, even in blight years, the death of the overground parts does not occur until late August, and about 10% to 15% of the crop is lost. If the blight is even later, and the plant dies in late September, the crop yield is hardly reduced at all, although tubers are lost in storage later.

2 Why do you think the time of infection affects the crop yield?

3 Why might the early varieties of potato escape blight more easily than the late varieties?

4 Apart from spraying a crop with a fungicide, how do you think potato blight might be prevented or controlled?

4.10 Cucumbers and costs

The growth of cucumber plants was investigated under controlled conditions in order to find the most economical way of producing the most, or the largest, or the earliest, cucumbers.

1 Why do you think anyone would be interested in finding out how to produce more, bigger, or earlier cucumbers?

Three factors were investigated: light intensity, temperature, carbon dioxide concentration. In all the investigations, the factor or factors investigated were kept at the level indicated from germination to harvest. The results are given in Table 4.6 as kilogrammes of marketable cucumbers.

2 Cucumbers are normally grown in glasshouses where the environment can be artificially controlled. The weather outside may be cold and overcast. Assume that conditions inside a glasshouse are at the minimum levels investigated: 15% light intensity, 20 °C and 0.03% CO_2. Which of the following conditions or pairs of conditions would you increase, and why, for maximum effect on production?

light intensity	temperature and light intensity
temperature	temperature and carbon dioxide
carbon dioxide	light intensity and carbon dioxide

Suppose you have a number of glasshouses, and cucumber is one of the plants you grow for sale to shops and markets. Your standard of living depends in part on the cucumbers you produce. Consider the exercise you would have to do before you decided whether to invest in lighting, heating and carbon dioxide production in your glasshouses. You would need to know the estimated costs of installing and running each. Then you would have to balance your investment against the expected return when you sold the cucumbers.

Suppose now that for the whole of the cucumber growing season, artificial lighting which directly affects the cucumbers costs £2x. Heating costs £3x. Carbon dioxide costs £x. Other costs are £x. You expect to get £x for each 10 kg of cucumbers.

Table 4.6 Yield of cucumber/kg

Light intensity necessary to give maximum photosynthesis/%	Temperature/°C	20		30	
	Carbon dioxide in the air/%	0.03	0.13	0.03	0.13
15		14	19	14	21
50		24	49	24	59
100		25	59	25	79

3 Would you boost all three factors, the one or two you selected in answer to question 2, or none?

4 Suppose, in addition, that increasing all three factors causes not only a higher total yield, but produces cucumbers which are ready for market some weeks before they would otherwise be. At this time you would expect to get twice the price for them: £2x for 10 kg. Would this cause you to change the decision you took in 3?

5 Finally, having taken the decision to invest heavily in extra apparatus for your glasshouses, how do you think you would boost the three factors?

4.11 Fish productivity

Oceans and seas cover more than three-quarters of the world's surface. But only 1% of the total world food production (10% of the world's animal protein production) comes from the sea.

On land, people stopped hunting and food gathering and turned to more settled ways of raising food in about 9000 BC. Instead of following the game with the next meal always unsure, animals were kept and reared in captivity. Goats were domesticated in what is now Iran in 8200 BC, sheep, pigs and cattle in Greece about 7200 BC, almost certainly imported as domestic animals from the Near East. One of the few remaining wild animals hunted by civilized man as a major source of food is fish. Others are whales, seals, some birds, molluscs, other types of sea food. Cows, poultry, even reindeer and frogs are kept, fed, then slaughtered at the appropriate time.

Table 4.7 Fish productivity

Ecosystem	Annual production of fish/g m^{-2}
Herbivorous and carnivorous fish together:	
World marine fishery (average)	0.2
North Sea	3.0
African Lakes	0.2 to 25
Carnivorous fish in fish ponds for angling (USA)	4.5 to 17
Herbivorous fish (carp) in fish ponds (Germany)	11.2 to 39.0
Carnivorous fish in angling ponds with added fertilizer	22.4 to 56.0
Herbivorous fish in ponds with added food and fertilizer	112 to 168
Ocean upwelling area, Peru (anchovies)	168

Table 4.1 in Problem 4.1 gives figures for annual net primary productivity of cultivated crops as high as 1200 to 4500 g m^{-2}. Compare Table 4.1 with the data given in the fish productivity table in this problem: Table 4.7. These latter data are for fresh mass and should be reduced by about half when making comparisons with the dry mass primary productivity of Table 4.1.

1 How do you account for

a the very large differences between the productivity noted in Table 4.7 and Table 4.1

b the differences in Table 4.7 between the yield of herbivorous and carnivorous fish

c the differences in Table 4.7 between the productivity of different areas?

2 Suggest ways in which fish could be treated as our other food animals are.

5

COMMUNITY ECOLOGY

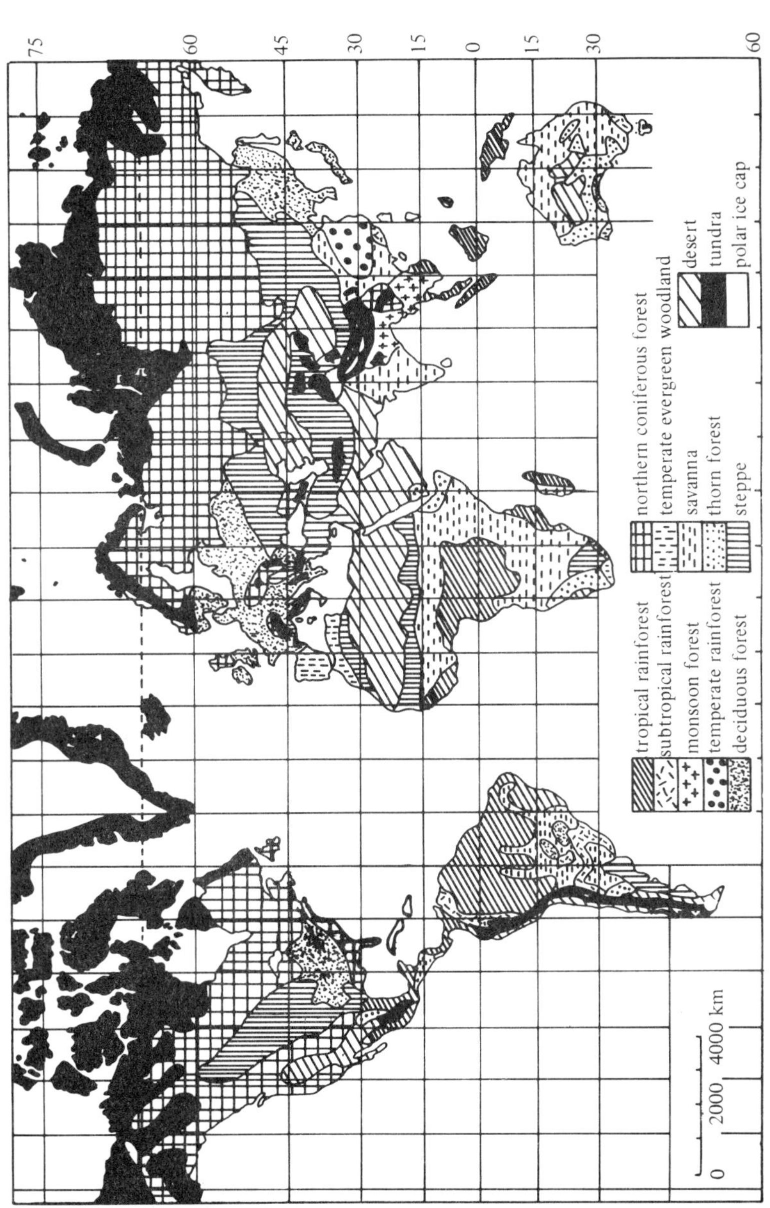

Fig 5.1 Distribution of the major types of world vegetation

5.1 Factors affecting ecosystems

1 Consider the whole of the British Isles, and then the other parts of the earth's surface you have heard about or seen (Fig 5.1). Make a list of all the distinct ecosystems you can think of.

Group your list under headings:
desert and tundra ecosystems
grassland ecosystems
forest ecosystems
marine ecosystems
fresh water ecosystems

2 Next, on a world-wide basis, consider all the factors which go towards determining what kind of community can live in an area. As just two examples, include the effect of grazing by herbivores and the effect of rainfall.

Group your list of factors under broad headings:

climatic factors

factors which relate to the activity of animals, including human activities (called **biotic factors**)

soil factors (called **edaphic factors**)

other **non-climatic physical factors** such as height above sea-level, slope and aspect.

3 Now put both lists together. For one or two ecosystems on your list, or for the communities shown in Figs 5.2 and 5.3, write down in order of importance the factors which you think were the most influential in leading to the establishment of that system. Alternatively, write down those which are the most important in maintaining it.

Fig 5.2 Mountain grassland at Cuzco in the Andes, Peru, at an altitude of 3600 m. The mean annual rainfall is 370 mm and the mean monthly temperatures range from 5 °C to 10 °C. The grasses are xerophytic and the animals are alpaca.

Fig 5.3 Oak woodland in Connemara near the west coast of Ireland. The mean annual rainfall is very high at 140 cm, and the mean monthly temperatures range only from 5 °C to 11 °C. The soil is thin, highly leached, moderately to strongly acid. The trees are short because of the low level of nutrients in the soil. Two layers are visible: the tree layer and the field layer consisting mainly of bracken and grasses, grazed by sheep.

For example, for tundra with permanently frozen water (permafrost) in the subsoil, temperature would be the most important factor, although others such as length of growing season may also be considered. For an ecosystem in water the volume of oxygen dissolved, the amount of light which can penetrate the water, the rate of flow and so on, are all important in determining what plants and animals can live there.

5.2 Succession and climax

The community which is characteristic of a particular area is called its **climax community**. A climax community is usually known by the name or names of its most common or **dominant** vegetation, as the plants are large and obvious and do not move, unlike the common animal species in the community. Of course, many climax communities are never achieved because of the activities of people. In the UK, for example, there are very few places where we have not influenced the vegetation to a considerable degree. The natural climax community of mixed deciduous woodland is therefore rarely achieved in this country in modern times.

If one starts with bare ground, then before the final community is reached a number of species will inhabit the area in stages. Such a progression of stages is called a **succession**, or **sere**. A succession is an orderly progression of community change, a sequence of communities each replacing the previous one.

1 Starting with similar places in the same area, do you think that succession would follow a similar course in each? For example, would a similar course be followed in each of a number of small ponds in a wood? Or in each field of an abandoned farm?

2 Do you think succession is one-directional?

3 If large trees are the dominant plant species in the final community, what stages in the sere might precede their establishment?

4 Do you think that the populations of successive communities replace the previous community gradually or quickly? Explain how the replacement might happen.

5 What do you think causes ecological succession?

6 When will a succession stop?

7 What influences the constitution of the final climax community?

8 Give examples of climax communities both in this country and in other parts of the world.

5.3 The number of species in a community

Different communities are made up of characteristic species, differing in type and number from other communities. Over the world's surface, some areas such as the tops of high mountains, deserts, and active volcanic islands have few or no species. Some, such as temperate upland areas, tundra, arctic waters, dunes and salt marsh have more species, but still relatively few. Other areas, such as the humid tropics, are teeming with different sorts of animals and plants.

Probably the most important factor which can be correlated with the number of species is the latitude. More species occur in tropical communities than in temperate ones, which themselves contain more species than are found in arctic communities. An example is species of ant. There may be only 10 species as far north as 60° latitude, up to 100 species at 40°, and up to 200 species within 20° of the equator. Other examples are given in Table 5.1.

Table 5.1 Number of species in three areas of North America

Latitude	Beetles	Snails	Fish near the coast	Flowering plants	Breeding birds
55°	170	25	80	400	50
40°	1900	100	200	1700	129
25°	4000	220	650	2500	295

1 We will consider first the implications for the community in the number of species which make up the community. Can you think of three effects which a high number of species in a community would have on the further development of that community? The effects relate to

a the exploitation of resources,
b variation in the total environment,
c ecological and evolutionary interaction.

2 What sort of major factors may be responsible for the increasing diversity of communities the nearer they are to the equator?

The list below outlines a number of hypotheses concerning the increased number of species in the tropics, as compared with temperate and arctic areas.

a The tropics are older-established than other parts of the world and have therefore had longer to develop. In addition, they have not had their flora and fauna disrupted by ice ages.

b Populations in tropical areas do not move at certain times of the year to a better climate. Thus there may be geographical isolation, causing speciation.

c Evolution may proceed faster in the tropics than elsewhere because there are more generations per unit time. Things grow more quickly with a faster return of materials to the environment on death. This allows increased selection. Predation, competition, parasitism and other biotic factors are more important than elsewhere, thus increasing selection and encouraging variation and adaptation.

d Species may not become extinct so easily in the tropics. There may be less competition, which could be due to more resources of all sorts, a more spread-out population for each species, and more control over competing populations by more predators.

e There are more rare species in the tropics because the environment is more stable. This is due to a more constant physical environment, and more stable communities.

f The tropical ecosystem is very complex and heterogeneous. There are, therefore, more niches than at other latitudes where the ecosystems are simpler.

g Primary productivity in the tropics is high due to the climate. More energy entering the community allows for more animal species.

3 Take each of the hypotheses listed above. Using your knowledge of ecology and evolution consider, by discussion, the extent to which you think each could be a reasonable hypothesis to account, in part, for the higher number of species in the tropics.

5.4 Rocky shores

Sea shores can be classified according to the type of substrate they possess, in and on which the organisms live. The substrate can be rock, sand, shingle or mud. The type of substrate has a major effect on the plants and animals which are found on the shore. Rocky shores offer many microhabitats and usually contain more species than the others. The stability of the rock means that it is less affected than the other three types of shore substrate by erosion. It also gives a firm and more-or-less permanent place where organisms can become attached. Sandy and muddy shores give no anchorage, and shift with the tides. Shingle is made up of large and small pebbles and also forms a very loose substrate.

Because of the more permanent nature of rocky shores, animals and plants can become established. There are usually clear and well-developed **zones** of organisms from the top of the shore near the land down towards low water mark. Each zone consists of its own typical animals and plants. Fig 5.4 shows zonation on an almost vertical rocky slope. There are lichens above the high water mark, then a zone of *Fucus*, then a wide band of *Ascophyllum*. Note that the pebbles at the base, which form an unstable substrate, are not colonized.

The data on p. 79 were collected from a rocky shore in spring. To collect the information, a ladder **transect** was made: a rope was stretched along the shore from dry land down to low water. The rope thus lay across all the zones. Every two metres along the rope, the species present in an area one metre square were recorded. For each species, an estimate was made of the frequency of that species in the square on a five-point scale:

1 = rare 2 = occasional 3 = frequent 4 = common 5 = abundant

At each recording place, the distance between a horizontal rope and the ground was measured so that a **profile** of the slope could be drawn.

Fig 5.4 Zonation of seaweeds on a rocky substrate

Table 5.2 lists the most common species which were found in the survey and their occurrence.

1 Present these data on squared paper. Plot the frequency variations for each species as frequency distributions along a line, as shown in Fig 5.5.

Draw the profile along the bottom of the sheet of paper so that the frequency in each square appears above the appropriate part of the profile.

2 Do there seem to be any main zones of plants or animals? If so, indicate them in some way in your diagram.

3 What factors do you think would cause the zoning? If you produce a long list of factors, try to arrange them in order of importance.

Table 5.2 Some species found on a rocky shore

	Distance from low water mark/m								
	Splash zone 16	HWM 14	12	10	8	6	4	2	LWM 0
Plants Lichens (especially) *Verrucaria*)	5	4	1						
Fucus spiralis		1	1	2	4	3	3	1	
Fucus vesiculosus			1	3	1	1	2	1	1
Fucus serratus						1	2	3	3
Laminaria spp								2	5
Animals *Littorina neritoides* (periwinkle)	4	4	4	3	2	2			
Balanus (barnacle)	1	2	3	5	4	5	5	3	1
Patella (limpet)			1	1	4	4	3	2	
Distance from rope to ground/m	0	1	2	2.5	4	4	6	7	7.5

Fig 5.5 Plotting the data

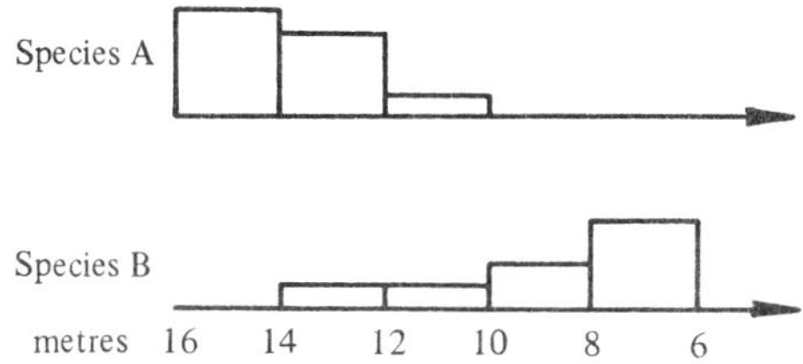

4 What sort of adaptations might plants or animals possess which enable them to survive the fairly rigorous and changing conditions on the upper shore?

5 Take any one of the adaptations you have mentioned. Suggest an experiment with the organism you choose which would enable you to determine whether or not the feature does indeed help it to survive.

5.5 Calcicole and calcifuge

Some species of plant are characteristically found in chalky areas or areas rich in lime and are called **calcicoles**. Some others, found in acid or lime-poor soils, are called **calcifuge** plants. Experiments with two species of bedstraw were set up to determine how well each species grew in each type of soil, either by itself or with the other species. The two species were slender bedstraw (*Galium pumilum*) which is a calcicole, and heath bedstraw (*G. saxatile*) which is a calcifuge.

1 If you were attempting to discover this

a what would be your experimental design,
b how would you measure the growth of the plants?

In this case, the experimenters set up a number of large pots containing either acid soil or chalky soil. In one-third of the pots with chalky soil, seeds of *G. pumilum* were planted and in another third *G. saxatile* was planted. Seeds of both were planted in the remaining pots.

The process was repeated with the pots of acid soil.

2 What results would you expect from the experiment described above?

In fact, the results were as follows. When they were grown alone both species grew well on both types of soil. When they were grown together however, the calcicole was the only one which was left in the chalky soil at the end of the experiment. The calcifuge was left on the acid soil.

3 What ecological principle does this demonstrate? (Refer to Problem 2.4.)

5.6 Earthworms in the soil

For one year, the number of earthworms in identical samples of soil was counted monthly. Each soil sample was one metre square and fifteen centimetres deep. The samples were taken from the same pasture land. The numbers are given in Table 5.3 together with air temperature at the time of sampling and mean rainfall for the whole month.

Table 5.3 Temperature, rainfall and number of earthworms for one year

Month	Jan	Feb	Mar	Ap	May	Jun	Jul	Aug	Sep	Oct	Nov	Dec
Number of worms	24	5	7	37	85	11	5	13	36	47	98	50
Temperature /°C	2	1	1	4	7	16	19	16	13	10	7	5
Rainfall /mm	40	30	25	50	80	20	5	25	40	50	80	70

1 Plot these data on one graph, using the left-hand vertical axis for numbers of earthworms and temperature, the right-hand vertical axis for rainfall.

2 What conclusions can you draw from the graph?

3 Both May and November have the same temperature and rainfall. Why do you think there are more earthworms in November than in May?

4 Suggest ways of improving the experimental design of this investigation.

5.7 Sampling

It is relatively easy to estimate the numbers and biomass of plants, but more difficult to do so for animals. For animals there are three basic methods which can be used.

A Taking the **total count** from a small area by, for example, placing a box over the area and removing in various ways all the animals trapped under it. The total number for a larger area could be estimated very roughly by multiplying up. Removing all the animals is, of course, a laborious exercise. Neither is it very accurate when trying to obtain an estimate of the population size of any one species.

B Removal sampling: To estimate the number of animals, say grasshoppers, in a larger area of about 100 to 200 square metres, the whole area is swept with a net. All the grasshoppers are removed and counted. The area is swept again, and any further grasshoppers similarly removed and counted. This is repeated. The numbers obtained from such a sampling technique are given in Table 5.4.

Table 5.4 Removal sampling

1 Sample	2 Number captured	3 Total number captured in previous samples
1	65	0
2	39	65
3	22	104
4	13	126
5	8	139

1 Plot two graphs: Column 1 against column 2
Column 2 against column 3

2 Compare these two graphs. Which graph enables you more easily to estimate the total population from the fewest samplings? From this graph, estimate the total population of grasshoppers.

C Capture, mark and recapture: A sample of animals is captured from the area under investigation. Each captured animal is marked and then released back into the population. A second sample is taken after the marked animals have had time to mix with the unmarked ones. The total population is estimated by using this equation:

$$\frac{\text{total population number}}{\text{total number marked and released}} = \frac{\text{total number recaptured}}{\text{number marked in recapture sample}}$$

Alternatively, the Haynes 'trap-out' calculation can be used. The number of unmarked animals caught each time is plotted on the vertical axis against the total number of animals previously caught and marked. The line is continued down to the horizontal axis to give an estimate of the total population.

3 Suppose 75 mammals had been marked and released. The second time, 60 individuals were captured, of which 48 were marked. Estimate the total population.

4 What assumptions are made in this procedure?

5 For each of the situations given below, say which of the three sampling methods given in this problem would be the most suitable one to use in order to estimate:

a the total number of animals in a piece of pasture land
b a woodlouse population in a garden
c the number of fish in a lake
d all the animal biomass in a small area for an energetics investigation
e the standing crop of consumers in field vegetation where animals drop to the ground or move away after disturbance and cannot be caught in a net
f the number of bank voles in a woodland.

5.8 Rat population sampling

This problem refers to the capture, mark and recapture sampling technique described in Problem 5.7.

An estimate of the number of rats in and around a wharf and warehouse was wanted before plans were made for their extermination. One evening, traps were baited and set between the wharf and the building. The following morning fifty-five rats were taken from the traps, marked, and released.

Two nights later the traps were set again. This time only twenty-four rats were trapped. Twenty of them were marked from the previous capture.

1 Use the equation given in Problem 5.7 to calculate the probable size of the population of rats in the area.

The same traps were set in the same place, on a third night, as well as some traps of a new design in a distant corner of the warehouse. That night, more rats were caught in the new traps than in the old ones. The rat-catchers thought of four possible reasons for the difference in numbers between the two.

a The new traps were more efficient than the old ones.
b Once it had been captured, a rat was suspicious of the traps and tended not to approach them.
c The activities of the investigators had caused an emigration of rats from the first area studied.
d The two sets of traps were sampling two different populations, one larger than the other.

2 Take each of the above reasons in turn and suggest how the investigators could determine whether or no it was a real reason for the difference in number of rats caught.

5.9 The Aswan High Dam

Egypt is mainly a land of sand with the River Nile running through it from south to north. The Nile has no tributary for the last 2700 km of its journey to the sea. Along its banks, there is a fertile green strip about six miles wide until the fan-shaped delta is reached.

Before the High Dam was built across the Nile at Aswan, the river flowed down through the Sudan and into Egypt. The High Dam now holds back an enormous reservoir called Lake Nasser. The lake has the river Nile as its base and is formed from the accumulated river waters. Lake Nasser stretches 400 kilometres back through Upper Egypt and into the Sudan. The High Dam is massive and a marvel of engineering. The river water roars through it at the side and the energy is harnessed in the hydro-electric station, generating enough electricity for home use and also some for export. There is no doubt that the High Dam has enriched the country. This problem is about some of the ecological side effects of such a massive project.

Until the Aswan dam was built, the river Nile had an annual flood in about August. The flood was caused by the rains in Ethiopia in the upper reaches of the Nile, which first caused the Blue Nile to flood before it became the Nile proper. By the farmers along the Nile the flood was awaited with some apprehension because they depended on it for irrigating and fertilizing the land. Too small a flood meant too little water. Flood water could be contained in channels and used for watering crops as long as there was any water left. As the flood receded from the land it had covered it left behind a fertile silt, brought down from the valleys of Ethiopia, which much lessened the need for artificial fertilizers.

Fig 5.6 Map of Egypt

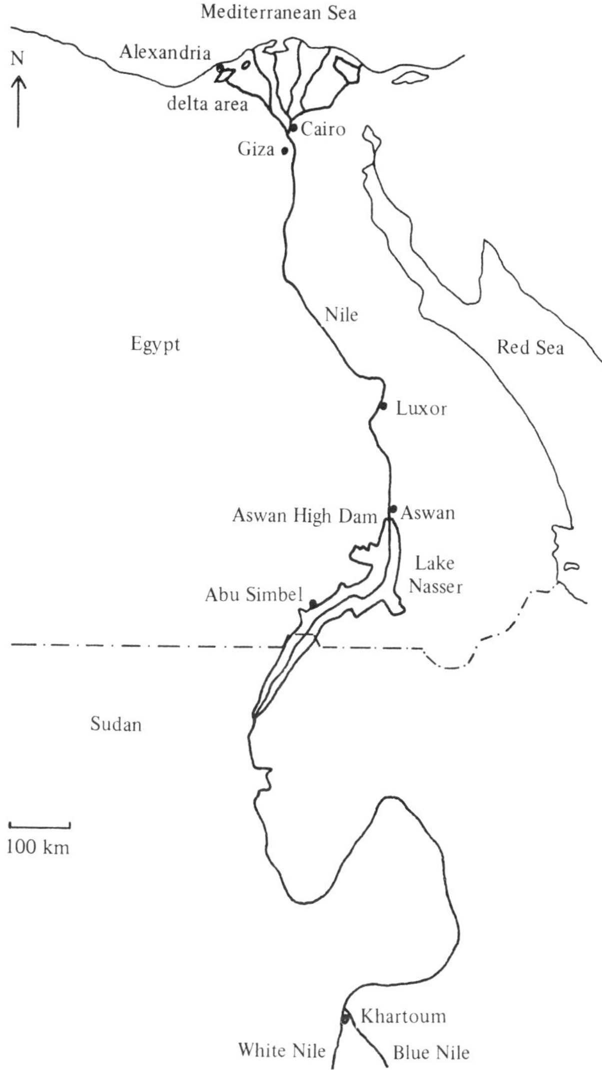

Some of the fertile silt was carried down to the sea and helped to provide food for the sardine shoals which lived there. In addition, the scouring action of the silt in the river helped to keep the river and delta clear, and the fresh water flooding out to sea prevented the encroachment of brackish water in the other direction.

The High Dam now regulates the river flow and there is no annual flood. But it has made permanent irrigation possible in the Nile valley, which makes it easier to grow crops. Permanent irrigation, however, means permanent breeding places for the desert locust. A number of parasites also live and breed more easily in permanent water than in water which comes and goes fairly quickly as the flood used to do.

Some examples, described briefly below, are the malaria-carrying mosquito; the Simulian flies which carry the cause of river blindness; and the blood fluke which causes bilharzia, sometimes called schistosomiasis.

The mosquito needs still water for its larva to develop, and the adult infects a person with the protozoan malarial parasite *Plasmodium* when it takes a blood meal.

The Simulian flies, which are small, blood-sucking blackflies, have a larval stage in water. When the adult bites a person it can transfer to man the parasitic worm which was present in its larval stage. The parasite is *Onchocerca volvulus*, a filarial worm. In man, it lives in connective tissues, especially those of the skin. The most serious effect of an infection is when the microfilarial larvae moving through the skin tissues reach the eye and are held up there. Blindness may result.

Schistosoma, the blood fluke, infects man directly when it is in its cercaria stage which leaves the intermediate host (a snail) and swims in water (Fig 5.7). When it meets the skin of a person washing, working or playing in the water, it breaks through the skin and enters the blood stream. The adult flukes live in the host's blood vessels. Eggs lodge in tissues and block up capillaries. They cause swelling and inflammation, damage to the liver, and death in heavy infections. If the host does not die, the parasite can live on for up to forty years. Bilharzia has long been a serious affliction, and there is evidence from mummies that the ancient Egyptians were similarly infected.

Fig 5.7 Life cycle of the blood fluke (*Schistosoma*) which causes bilharzia

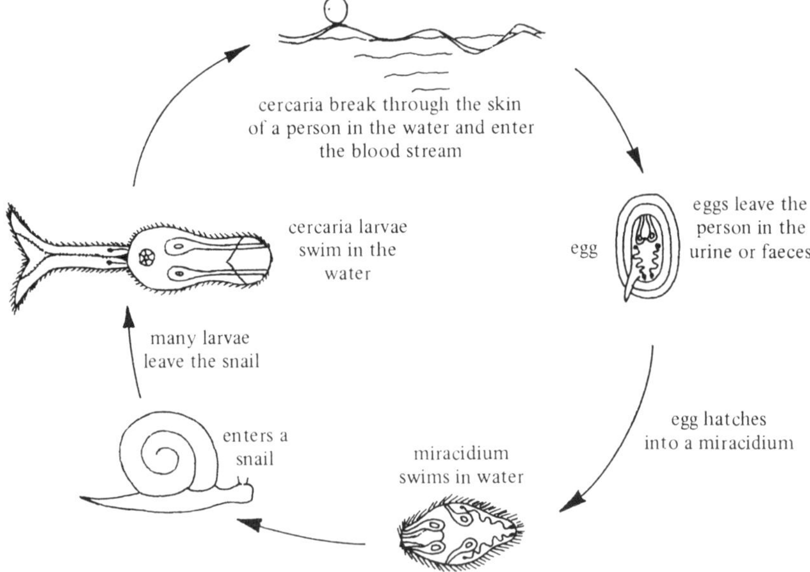

A more direct effect of the building of the High Dam was the moving and re-housing of tens of thousands of people who lived on land which would be covered by the lake formed behind the dam.

In addition, many monuments from the time of Ancient Egypt were to be covered by the water, and lost for ever. The religious beliefs of the ancient Egyptians, from about 4000 years BC, involved the burying of the dead with elaborate mummification and with their property around them. Because of the hot dry climate a great deal of this evidence of the past was preserved in the sand. This was unlike, for example, the early civilizations of the Rivers Tigris and Euphrates where environmental conditions rapidly caused the decay of the evidence. In Egypt, even ancient unmummified bodies in shallow graves have been discovered with some degree of preservation of the flesh. In 1976 the mummified body of a singer called Waty, who died 4350 years before, was discovered wrapped in linen bandages in a coffin. It is the oldest known mummy in the world.

The area to be covered by Lake Nasser contained a large number of monuments of one sort or another. Before the lake water rose, a number of temples and other archaeological sites were investigated as fully as time allowed. Many were not, and no doubt many more were even then undiscovered. One of the most famous sites, however, at Abu Simbel (Fig 5.8), was raised from the ascending waters of the lake in a complex and costly operation, paid for by cash from many countries. The two temples at Abu Simbel, built by Rameses II who was Pharaoh from 1298 to 1232 BC, were cut out of the rock in huge blocks and reassembled at a higher level. They now overlook the lake just as they looked east across the Nile for over 3000 years, and the rising sun still strikes the innermost altar with its first rays. Yet another famous monument, the temples of Philae at Aswan itself and situated just behind the dam, are being lifted from the lake and placed on higher ground. Again, money for this is coming from all over the world. That raised by the exhibition of the treasures of Tutankhamen's tomb at the British Museum in 1972 went towards the Philae operation.

Fig 5.8 The main temple at Abu Simbel

This brief outline indicates some of the many implications, a large number of them ecological, which follow a decision to change to such an extent the natural course of events. Imagine and consider carefully the situation as it used to be and as it is now, and then write brief notes in answer to the following questions.

1 What would be the reasons for wanting to build a dam?

2 Do you think the advantages of having the dam continue unabated into the future?

3 What seem to have been the disadvantages of the dam?

4 If you have time, look up and make notes on the life histories of the parasites mentioned, and say how they could be controlled.

5.10 Life on Mars?

One of the tests for life on Mars used by the Viking spacecraft in August 1976 utilized radioactive carbon dioxide ($^{14}CO_2$). For five days Martian soil was exposed to the Martian atmosphere to which traces of $^{14}CO_2$ (and labelled carbon monoxide) had been added, while a bright lamp lit the sample. At the end of five days, the soil was heated to 1100 °C. Any organic matter present would be broken down with the soil and vaporised. The resulting vapours were passed through an apparatus which could detect radioactive carbon. The detector found more than six times the amount of ^{14}C than would have been expected if the soil had been sterile.

1 What earth process was this test looking for?

2 Two control experiments were done. What do you think they were?

In the second test, called the labelled release experiment, a nutrient soup containing radioactive carbon (^{14}C) was incubated with the soil. Thereafter, the amount of $^{14}CO_2$ in the atmosphere above the Martian soil was recorded. The results are given in Fig 5.9, together with earth and lunar soils for comparison. The control soil was heated to 160 °C for three hours before the injection of the nutrient.

Fig 5.9 Results from the labelled release experiment on raw and sterilized samples of Martian, Lunar and Antarctic soil

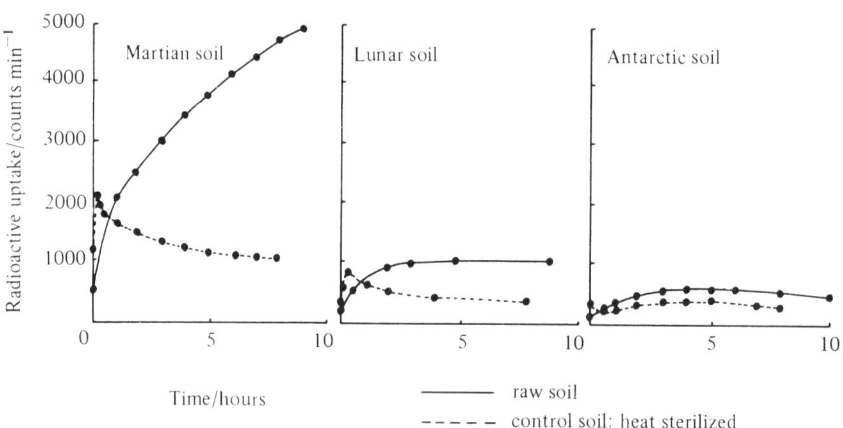

3 What earth process was this test looking for?

4 Do you think it is possible to conclude from these data that there is life on Mars?

One of the scientists analysing these results said that if he had been using earth soil and had got results like this he would have concluded that life was present, that there was a "weak positive biological signal". Mars, however, may be different. If one does not hypothesise life, then there is at least some fascinating chemistry on Mars.

6

POLLUTION

6.1 Freshwater pollution

Rivers and streams may be polluted in a number of ways. They may be polluted by

a domestic pollution, where sewage or other organic matter is put into the water,
b soil pollution, where excessive silt from land erosion or poor soil use runs into the water,
c industrial or chemical pollution, when toxic substances or organic waste products such as those from food or paper mills, textile manufacturers, laundries, breweries, cattle yards and slaughterhouses (to name only a few) are discharged,
d heat, such as the warm water from the cooling towers of power stations, or other industrial effluents, which raises the water temperature.

The animals which live in fresh water are mostly intolerant of a reduced oxygen supply. Organic material, whether it is sewage from **a** above or organic waste products as in **c**, decomposes in the water. It is the oxygen demand of the decomposer organisms which causes the low concentration of dissolved oxygen where there is too much organic matter. The decomposers (mainly bacteria) involved in the decay use oxygen, and thus reduce the amount available for the freshwater animals. If the oxygen demand is high, the concentration of dissolved oxygen in the water falls, and animals and plants die from lack of oxygen. The freshwater animals cannot return to the stretch of polluted water until the oxygen level returns to normal. Warm water (**d**) contains less oxygen than cold, with the same effects. **Biological oxygen demand (BOD)** is a term which means the amount of oxygen which can be consumed by the oxygen users present in the water. It is a measure of the activity of micro-organisms in decomposing organic matter.

Some of the ways in which pollution may be detected and measured are:

i By the presence of an indicator organism. *Escherichia coli* which is present in faecal matter can be detected by the coliform test. *E. coli* itself is a non-pathogenic bacterium which lives in the intestine. It is easier to detect, and longer living, than pathogenic bacteria. If it is found, however, there is a chance that some pathogenic bacteria are also present and that both have come from faecal matter.
ii By the effect on water animals and plants. Unpolluted water has a characteristic fauna and flora. The structure of the community changes when the water is polluted.
iii By chemical tests to determine the amount of oxygen dissolved in the water.
iv By chemical tests for specific polluting materials.

This problem is about the effect of organic pollution on the freshwater community. As one goes downstream from where the organic matter is discharged into the river and the chemicals and other materials in the water change, the flora and fauna also change (Fig 6.1). When the volume of dissolved oxygen is reduced the number of non-tolerant species becomes much less. Not all animal and plant life disappears, however, as some species are tolerant. Their populations may even increase. Primary productivity may also increase with the addition of rich sewage material, so total numbers of individuals may be high. With very severe pollution, even the number of tolerant species is reduced. As the water returns to normal, some distance from where the polluting material is put in, the usual flora and fauna reappear.

Fig 6.1 Freshwater animals associated with organic pollution

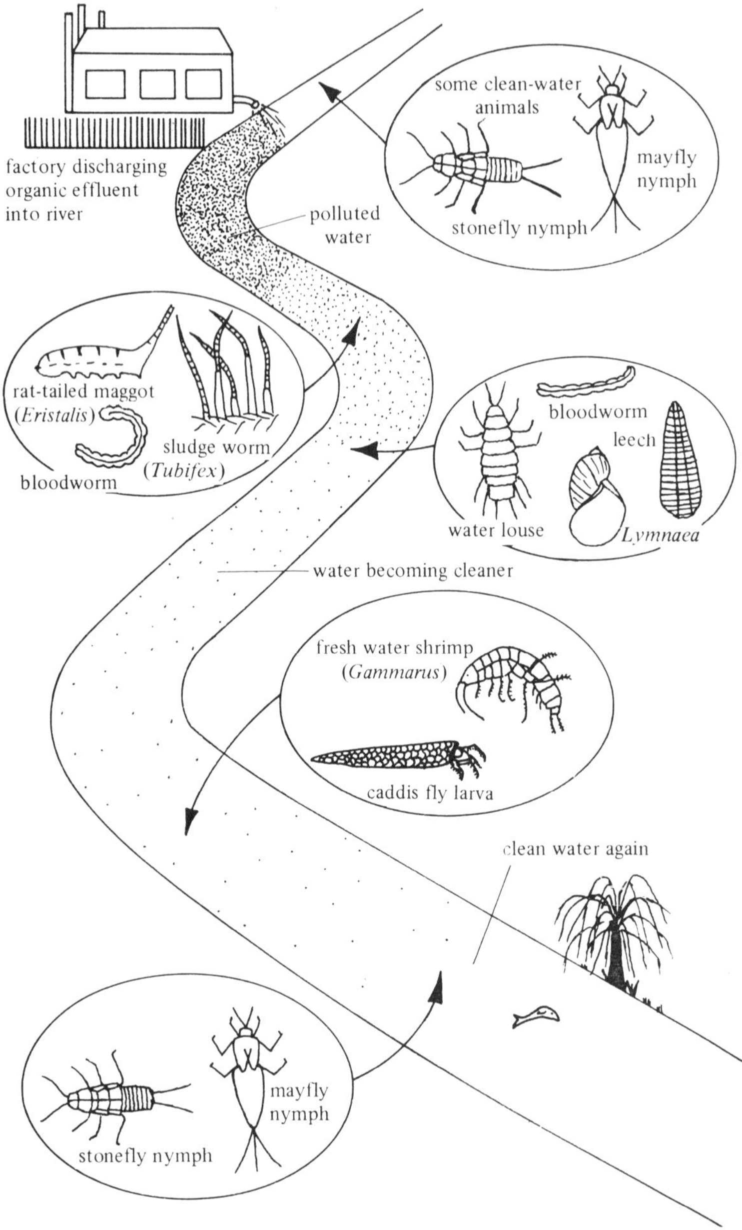

1 Look first at Fig 6.2. It shows diagrammatically what happens to the BOD, the dissolved oxygen, and the suspended solids when sewage or other organic material is added to a river or stream.

a Say which curve is which.

b Briefly describe what these three characteristics of the water are like before the sewage is added. Explain the reasons for the changes in the curves until the levels become normal again.

Fig 6.2 The effect of organic pollution on fresh water

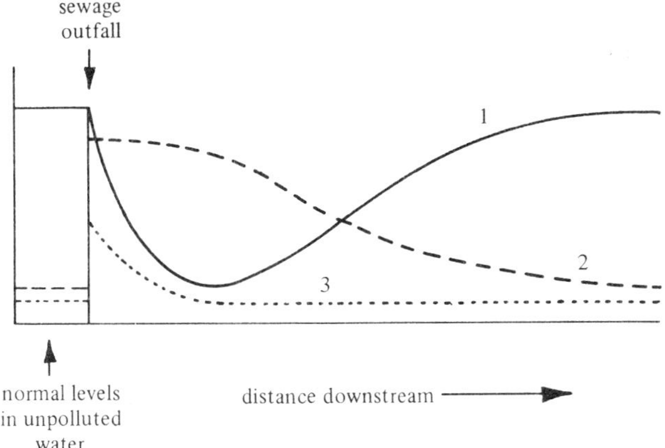

2 Fig 6.3 shows what happens when hot water is added to the river shortly after the sewage.

a Again, label the two curves, and give reasons for your answer.

b Local river boards set a maximum limit for BOD in factory effluent. The maximum is higher in winter than in summer. Why?

Fig 6.3 The effect of hot water on an organically polluted river

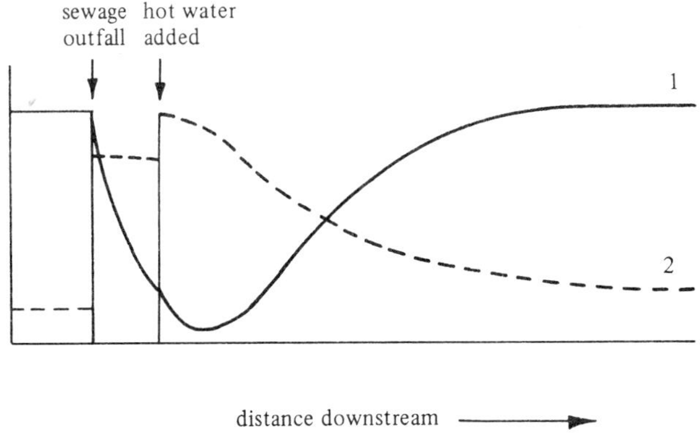

Fig 6.4 shows the variation in some chemicals and in key organisms after sewage or other organic matter has been added. The sewage fungus shown in B can be a mix-

ture of various decomposer organisms growing on the stream bed: bacteria, fungi, protozoa, rotifers, nematodes and chironomid larvae. *Cladophora* is *C. glomerata*, a filamentous green alga called blanket weed.

Fig 6.4 Other effects of organic pollution on fresh water

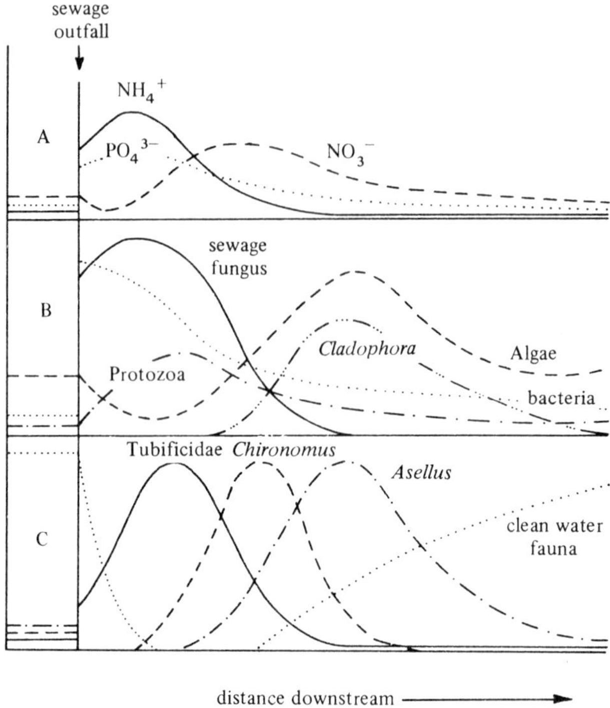

3 Fig 6.4A shows ammonia and phosphates released from the decomposing organic matter. As oxidation proceeds the ammonium ion (NH_4^+) is converted to nitrate (NO_3^-). As the concentration of ammonium ions drops, the concentration of nitrate ions therefore rises. Relate Fig 6.4A to the growth of algae in Fig 6.4B.

4 Fig 6.4 is drawn to the same horizontal scale as Fig 6.2. To what is the re-appearance of the clean-water fauna related?

5 The Tubificidae (tubifex worms), *Chironomus* (bloodworms), and *Asellus* (water louse) are pollution fauna which succeed each other downstream. Which is the most tolerant of pollution?

6 What outside factors do you think would shorten the length of the polluted part of a river and hasten its return to normal?

7 It looks, then, as though some organisms or communities may be taken as **indicators** of pollution. What use do you think can be made of this information?

8 Discuss what would happen in a lake if similar polluting materials to those described above were discharged into it.

6.2 Air pollution

The substances which pollute the air include smoke, sulphur dioxide (formed when coal burns from the sulphur in coal), deposited matter such as grit and dust, and exhaust products such as carbon monoxide and hydrocarbons from motor vehicles. The first three are the ones most commonly measured in Britain. Although rising to high levels this century, air pollution is not something new. Even in thirteenth-century England it was recognised as a public nuisance needing some kind of legislation, and down the centuries city dwellers have complained of it.

Air pollution can kill plants or reduce their growth and the yield of crops. It affects animals, and can cause illness and hasten death in man. The London smog (a sulphur type) which started on December 5th 1952 became so dense that by December 7th visibility was reduced to one metre in places. One or two people were known to walk accidentally into the river. A party sent out to search for an aeroplane which had taxied to the wrong place at the airport got lost. The smog also caused the premature deaths of about 4000 people in one week. It is mainly old people and those with respiratory diseases, the new-born and the frail, who succumb. Indeed, the bronchitis patients in London hospitals noted discomfort some six to twelve hours before others could detect that smog was at hand.

As in water pollution (Problem 6.1) there are **indicator organisms**–apart from bronchitis patients–which can be used to detect pollution. A well-known investigation was done by O. L. Gilbert into the distribution of lichens in the Tyne Valley. The area around Newcastle at the mouth of the River Tyne in north-east England is an industrial and mining area with a long history of air pollution. Most of the pollution came from the burning of coal by industry and in private houses. The built-up area studied was about 13 km across, with small towns also extending up and down the coast for many kilometres. The north and north-west of the area, however, was open country with no large towns. In addition, the prevailing wind blew from the country towards the industrial area (Fig 6.5).

The investigation mapped lichens growing on three substrates: tree trunks, sandstone, asbestos (e.g. asbestos roofs). Ten sites along a 16 km transect running north-west from the city centre were also carefully examined. Figs 6.5 and 6.6 show the results. In the centre of the area only five species of lichen were found. If a healthy lichen and the substrate on which it is growing is taken from the country to a city centre, it is the algal partner which dies first.

1 What conclusions can you draw from Fig 6.5?

2 Sulphur dioxide is most toxic under acid conditions. As the pH rises and conditions become less acid, sulphur dioxide is oxidised to sulphite ions (SO_3^{2-}), and then to sulphate ions (SO_4^{2-}) which is non-toxic at the concentrations found in Newcastle. Ash was the most common tree in the area and ash trees tend to provide an acid habitat. It is thought that it is the sulphur dioxide in the air which causes the death of lichens. Relate these facts to Fig 6.6.

There are similar lichen deserts in the centre of most large towns, followed by an area of transition to open country. In some parts of the world, a city is so situated that something called **thermal inversion** occurs over it. Los Angeles in the United States or Santiago in Chile, South America, are good examples. Both these cities have mountains at one side and sea nearby at the other. Both have a great deal of motor traffic and bright sunlight. Normally the warm surface air over a city will rise. Inversion sets in when cool air from the nearby sea moves in under warm air, there is low wind, high atmospheric pressure, and therefore a stable air column. The normal temperature gradient is reversed by the cool air in and on top of the city, and the warm air above that.

Fig 6.5 Lichens in the Tyne Valley

Wind direction

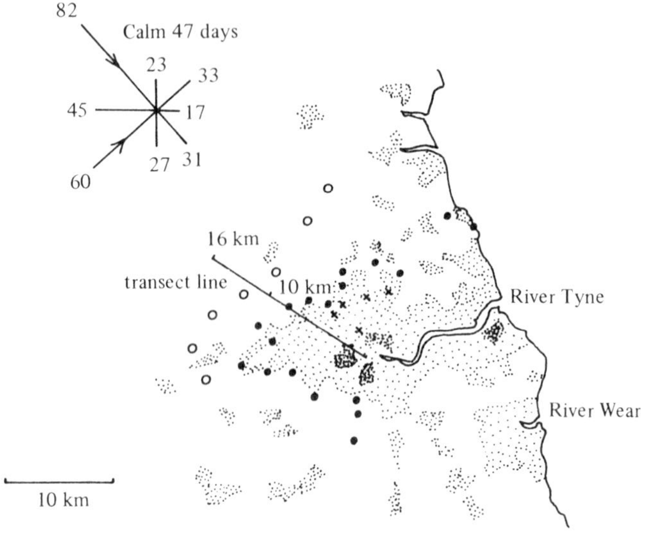

o limit of lichens on ash trees

● limit of lichens on sandstone walls

× lichens recorded on asbestos roofs

built-up areas

high density housing

Fig 6.6 Number of species of lichens on different substrates along a transect from the city centre

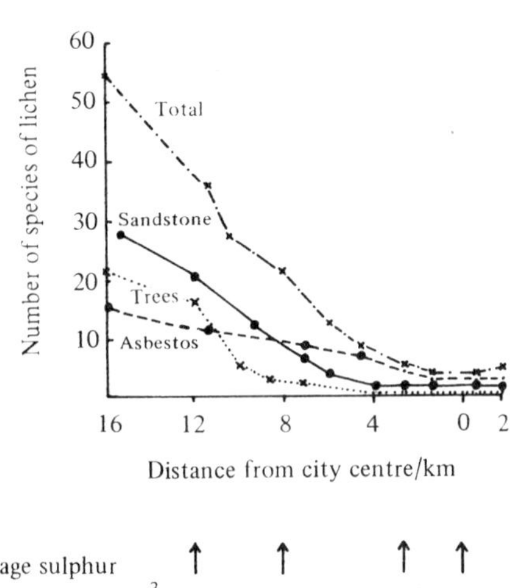

Annual average sulphur dioxide concentration/$\mu g\ m^{-3}$: 65 90 160 200

3 Why do you think that thermal inversion will cause a smog?

4 What measures have been, and could be, taken to lessen air pollution?

6.3 Bronchitis

Bronchitis is the main cause of death in men over 45 years in the UK, and the fourth leading cause of death for the population as a whole. In addition it causes the loss of over thirty million working days a year. It is a disease which seems to be associated with air pollution.

People suffering from bronchitis have a persistent cough and breathlessness. They find it progressively less easy to breathe. The effect is caused by the narrowing of the smaller branches of the bronchioles. Emphysema is also a part of the disease: several alveoli merge into one larger sac, which has less surface area than the greater number of small alveoli and thus can absorb less oxygen (Fig 6.7). Infection of lungs and bronchi is easier than it is when they are functioning normally. Eventually, when the emphysema has become advanced, the heart has a heavy job to pump the same volume of blood round far fewer capillaries in the lungs. The heart enlarges and there may eventually be heart failure.

Fig 6.7 Normal and diseased alveoli

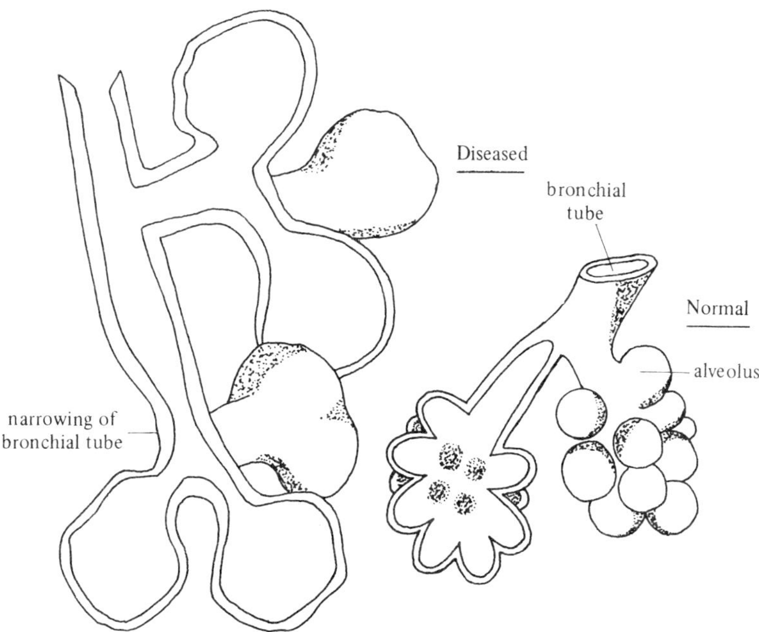

The lungs and bronchi are, of course, exposed to the air we breathe. They are delicate organs and it is not surprising that they react to pollutants. The larger the city, the more bronchitics there are. Post Office employees who work indooors have less bronchitis than postmen who deliver letters. Postmen who work in the centre and north-east of London have a bronchitis rate nearly twice that of the postmen in the less polluted areas of London.

Fig 6.8 shows the standardized mortality ratio for men who died from bronchitis in an industrialized part of the country: Newcastle and Tyneside. A figure of, say, 200 for the standardized mortality ratio for an area means twice as many deaths from that cause as is normal (100) for the whole country. The area shown is the same area as that shown in Fig 6.5 in Problem 6.2. The UK Medical Research Council's air pollution unit has shown that the condition of bronchitis sufferers gets worse when 24-hour averages for smoke and sulphur dioxide rise over 250 microgrammes per cubic metre for smoke and above 500 microgrammes per cubic metre for sulphur dioxide.

Fig 6.8 Deaths on Tyneside due to bronchitis

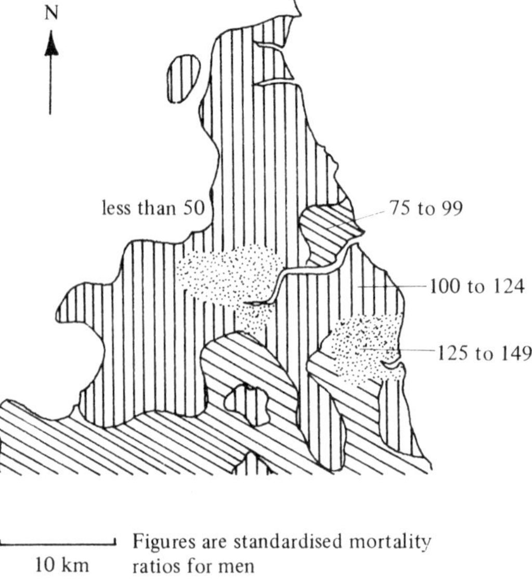

Fig 6.9 Bronchitis and smoking

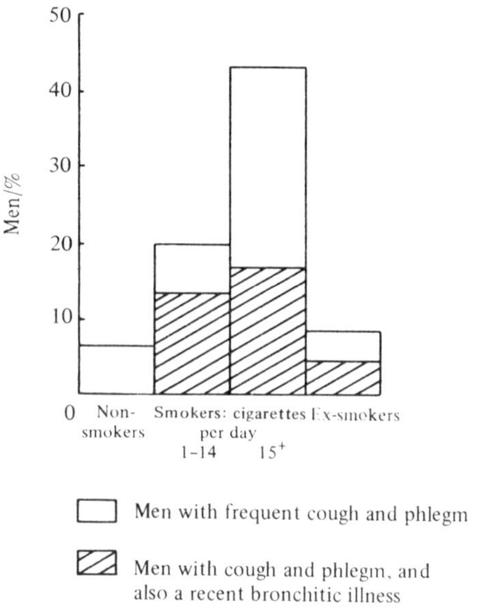

Fig 6.9 shows bronchitis in men aged between 55 and 64 years related to an air pollutant which is directly under the control of the individual: cigarette smoke. The 'recent bronchitic illness' had to be one which caused loss of work for a week or longer in the previous three years.

Consider the data in these figures. Discuss the relationship of bronchitis to air pollution and consider ways of preventing bronchitis.

6.4 Copper tolerance

Mining for substances such as lead, tin or copper, may produce large mounds of earth called waste tips, rich in metal compounds. Few species grow on such waste tips because of the toxicity of the soil. But a few individual plants of a few species, tolerant to the toxicity, can grow there. They give a characteristic look to the flora of waste tips, and have been much investigated.

The few tolerant species which grow on such waste tips grow also in normal soil, where they occur with many other species. One possible hypothesis to account for the fact that the tolerant species can survive on waste tips is that all individuals of the species are tolerant to the chemical concerned. Another hypothesis is that only where they grow on toxic soils can they tolerate the toxicity. In other words, tolerant individuals of the species arise by natural selection.

1 Design experiments to investigate each of the two hypotheses given above.

2 How do you think one could determine if a plant is tolerant or not?

In fact, it has been found that it is root growth which is mainly affected by metals in the soil. The roots of tolerant plants can grow very much better in polluted soil than can the roots of non-tolerant plants. Tolerance can be measured by growing plants in (for example) a copper solution of 0.5 ppm (0.5 mg dm^{-3}) and a normal culture solution. The tolerance can be expressed as an index: the length of root in the toxic solution compared with root length in a normal solution. Thus, if a plant has 10 cm roots in a copper solution, and 20 cm in a normal solution, it has a **tolerance index** of 50. Tolerant plants continue root growth in the toxic solution to an extent which depends on how tolerant they are. Non-tolerant plants do not continue root growth.

One tolerant species, *Agrostis tenuis*, is a grass. It is wind pollinated, and is unable to fertilize itself (**self-incompatible**). It grows in normal pastures, and it also grows across a waste tip of a copper mine in Wales, showing copper tolerance.

Table 6.1 shows the percentage survival of normal and tolerant plants when seeds from each were grown in different soils.

Table 6.1 Survival of *Agrostis tenuis* seven weeks after germination in three types of soil/%

	Potting compost	Normal soil	Toxic soil
Seeds taken from normal plants	41	33	0
Seeds taken from tolerant plants	35	41	100

Fig 6.10 shows the tolerance of adult populations of the grass growing in the soil on either side of the tip and across the tip.

Fig 6.10 Copper tolerance of adult populations of *Agrostis tenuis* across a copper mine

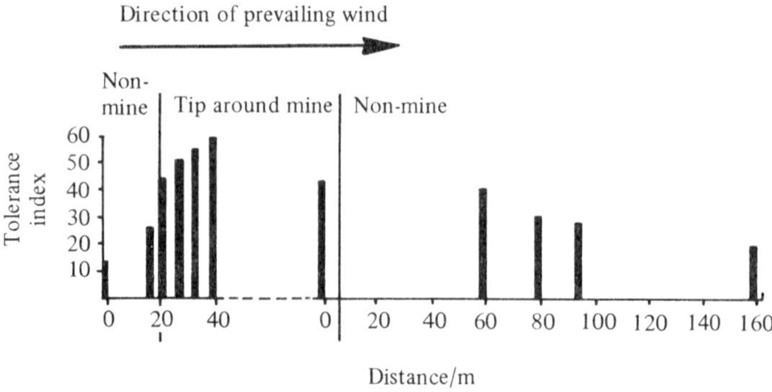

3 Suggest explanations for

a the presence of tolerant plants in the non-toxic soil around the tip

b the high tolerance index shown by plants which grow in the toxic soil.

4 Using the table and Fig 6.10, say what evidence supports the hypothesis that there has been selection, and when you think it occurs.

5 Fig 6.11 shows the tolerance of the same adult population already shown in Fig 6.10, plus the tolerance of the populations along the same transect which have arisen from seed produced by the plants. Suggest a reason for the fact that the seed populations in the ground outside the mine area at the right-hand side of Fig 6.11 are more tolerant than the adult populations by their side, whereas those on ground inside the mine area are less tolerant than the adult populations.

Fig 6.11 Copper tolerance of adult and seed populations of *Agrostis tenuis* across a copper mine

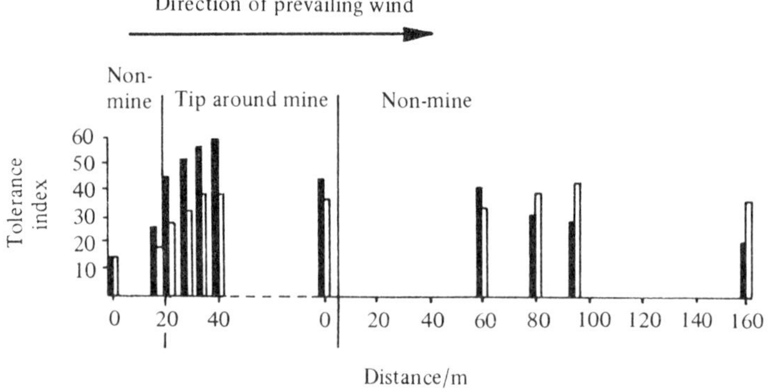

Environments which are so hostile that only a few characteristic species can live there are useful to the biologist for a study both of ecology and of evolution. Such environments are usually very easily recognized ecologically and spatially. One physical factor usually dominates them. It is relatively easy to determine the success of a species which occurs there in competition with other species, and to see variation in tolerance amongst the offspring of tolerant individuals. Evolution in extreme environments can perhaps be seen as part of the normal process of evolution, as any species which evolves in a new situation will be doing so in a situation which, for it, is somewhat hostile. For the biologist an extreme environment provides an opportunity to observe and experiment.

6 Name some other hostile environments occupied by a characteristic flora and fauna, and give the dominant physical factor.

6.5 Pesticides

A **pesticide** is a chemical which will attack and kill a pest. A pesticide designed to kill insect pests is called an **insecticide. Molluscicides** kill mollusc pests, a **nematicide** kills nematode worms, a **herbicide** is used to attack weeds, a **fungicide** for a fungal infection and so on.

There have always been animal pests of food crops, with heavy losses due to their activities, both while the crop was growing and when it was stored after harvest. Pesticides of some sort have been used for centuries. Pliny in the first century AD used arsenic, nicotine was used from the seventeenth century, soap and sulphur from the 1820s,Bordeaux mixture in the last part of last century was based on copper(II) sulphate, and so on. Large scale manufacture and use of pesticides, however, began around the middle of this century.

Fig 6.12 Diagram to show the development of pest control and pest fluctuations

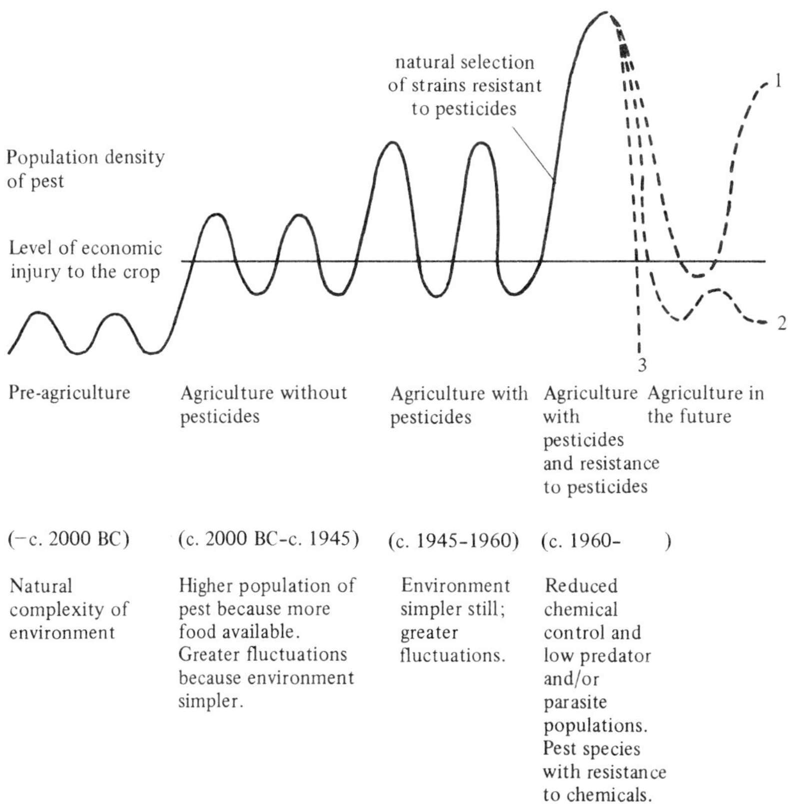

Fig 6.12 shows a diagrammatic representation of the historical development of pest control and past fluctuations. In the time before agriculture as we would recognize it, pest populations would be at a low level. The animals would hardly be pests as the natural complexity of the ecosystem did not allow huge populations to build up. In natural communities most species of animal do not reach the abundance they can reach on a crop. Then, as more food became available, higher populations of pests were possible on crop plants. But the environment was simple, and it was easy for a population to crash. The population curve of a pest therefore fluctuated.

When pesticides were used the population fluctuations became larger because the environment was even simpler. Natural enemies of the pest, as well as the pest itself, were removed by the pesticide. Without its predators, the pest could then build up a bigger population later from individuals which had escaped the toxin, or had immigrated from another area. Following the appearance of pest strains resistant to pesticides, pest populations could become even larger.

N. W. Moore, in the figure (Fig 6.12) reproduced here, forecasts three possible future directions of the curve shown. The direction of the curve could be affected by

a integrated control, where some pesticides are used together with biological control and some control by normal agricultural practices
b artificial environments, closed to the outside and from which pests can be excluded
c the use of new pesticides.

These three forms of control correspond to three characteristics of the environment:

i an increased simplification of the environment, with total control
ii the continuation of a fairly simple environment, with short-lived control
iii a partial return to complexity of the environment, using the complexity to control.

1 Match these two lists to each other.

2 Look carefully at Fig 6.12. Match each of **a**, **b** and **c** above to one of the three dotted curves.

6.6 Insecticides

Chemicals which kill insects (**insecticides**) are widely used over the whole world. One of the best known, and longest used, is **DDT**. It was first used on a massive scale about the mid-1940s. It and its effects have been thoroughly studied, and its ecology is now well known.

As a killer of insects DDT is extremely effective.

a It is used to reduce the number of malaria-carrying mosquitoes. As a result millions of lives have been saved. There are now about 120 million cases of malaria per year throughout the world, as compared with 300 million cases (3 million deaths) per year before 1946. Anti-malaria campaigns are now the major consumers of pesticides. 27 000 tonnes of DDT are used each year, plus about 30 000 tonnes of other pesticides.
b When DDT was first introduced in Sri Lanka, the adult death rate fell by 66%, and malaria was completely eradicated from the island.
c DDT has also markedly reduced the incidence of other insect-carried diseases such as typhus, yellow fever, sleeping sickness (trypanosomiasis, caused by a protozoan parasite carried by tsetse flies) and river blindness (onchocerciasis, caused by a nematode which is carried by the blackfly; in moving through the

connective tissues of its human host, the nematode parasites may reach the eye and cause blindness).

d Killing insect pests of crops greatly increases the yield of marketable produce. In addition, the vegetables or whatever is being marketed look better with no insects and no insect damage.

On the other hand, DDT has other properties or side effects, which have caused it to be banned from widespread and indiscriminate use.

i In some cases the use of insecticides has gone wrong where useful animals such as bees which pollinate fruit trees and crops have also been killed.

ii In other cases the pest increased enormously the year after spraying because of the reduction of natural predators and parasites. They were either also killed by the insecticide, or else they depended on the density of their prey to keep up their numbers.

iii Sometimes reducing the numbers of a particular pest has caused another, more resistant, pest to increase.

iv Strains of many insects resistant to specific pesticides have evolved.

v All animals tested have some DDT in them. The milk from mothers who feed their new-born babies contains DDT and no human body tested has been free of DDT. Even snow from the Arctic and Antarctic (except in the immediate vicinity of the South Pole) contains DDT, and air 8500 metres up contains traces of it.

vi DDT accumulates at successive steps in a food chain.

vii DDT residues are found in all foods.

viii Evidence shows that DDT reduces the reproductive rate of birds to a level which could cause the extinction of rare species.

ix The long-term effect of DDT on man or other animals is not known.

x DDT is very resistant to decomposition. It stays in the soil for many years. It is chemically stable and stays poisonous for a long time.

Many studies have been done of animals and people exposed to insecticides. One study, sometimes used to support the continued use of DDT, was of workers in a DDT factory. The evidence given to support the claim that DDT was harmless to people was that over two years the workers in the factory showed no ill-effects from heavy exposure to DDT.

1 Criticize the experimental design of this study.

2 How do you think DDT gets into the snow around the poles, or air 8500 metres up?

3 Discuss the problems raised in the pros and cons of DDT listed above. Start by asking if the benefits outweigh the dangers. Consider possible long-term effects, and other ways of pest control.

4 If you were responsible for the work of a team of research scientists trying to synthesize a new insecticide, what criteria, and in what order, would you lay down for the finished product? Consider such points as its specificity, cost, persistence and breakdown.